2009 YEARBOOK

edited by

Jonathan and Corinna Downes

Typeset by Jonathan Downes,
Cover and Layout by Spid-ohPU$$$Y for CFZ Communications
Using Microsoft Word 2000, Microsoft , Publisher 2000, Adobe Photoshop CS.

Photographs © 2009 CFZ except where noted

First published in Great Britain by CFZ Press

**CFZ Press
Myrtle Cottage
Woolsery
Bideford
North Devon
EX39 5QR**

© CFZ MMIX

All rights reserved. Without limiting the rights under copyright reserved above, no part of this publication may be reproduced, stored in or introduced into a retrieval system, or transmitted, in any form of by any means (electronic, mechanical, photocopying, recording or otherwise), without the prior written permission of both the copyright owners and the publishers of this book.

ISBN: 978-1-905723-37-9

CONTENTS

5. Introduction
7. Statement of Core Belief
9. *In the wake of the winged sea serpent* by Theo Paijmans
21. *Star Rot - Rot from the stars? Or a load of old rot* by Jon Downes
29. *The Tarasque* by Richard Muirhead
35. Prologue to Michael Woodley's article
39. *Human diversity and hominology* by Michael Woodley
61. *The mystery cryptic animal of northern New South Wales* by Gary Opit
81. *Folkloric creatures of Philadelphia and the immediate counties* by Neil Arnold
91. *Strange animals in Snowdonia* by Oll Lewis
113. *The Black Tapir of Brevet - a Malayan Mystery Beast Resurrected* by Dr Karl Shuker
119. *Spiritualism and UFO Mythology* by Dr David Waldron
129. *Monstrous Montana* by Michael Newton
137. *Bellissimo Beasts: Cryptozoology Italia* by Neil Arnold
155. *`Jack` the Horse Ripper* by Jon Downes
163. *Telegram from Oz: 2008 at the Australian branch of the CFZ* by Ruby Lang
165. *CFZ:USA 2008* by Nick Redfern
173. CFZ Annual Report 2008
181. CFZ: 2008 - the year in pictures

INTRODUCTION

It has been over 14 years since I first came up with the idea of the CFZ yearbook. The volume that you are reading is the 11th of the series. I think that is pretty good going.

The CFZ has changed a lot in those 14 years. We have gone from being a bunch of amateurs who really didn't know what we were doing, but did it for fun, to being a team of dedicated professionals doing our best to change the world just that little bit. However the CFZ has changed even more in recent years. We have always been aware of the injustices and failings of the societies in which we live, but recently we have discovered that the work we do can actually be challenged into socially constructive projects, and as a result, we have become far more political.

Ironically, one of the reasons that we have had to become more political is because other people have begun to bring politics, and the politics of religion, into cryptozoology and its allied disciplines. No; that is actually slightly misleading - other people have been bringing politics, and the politics of religion, into cryptozoology for decades, and we have done nothing about it except to grumble. Now, as the first decade of the 21st century approaches its end, we have decided to do something about it. For the last 14 years we believed that it was enough to investigate the mysteries of the natural world, and then to publish our findings. Now we believe that we have to do more.

If the CFZ is to be anything more than a bunch of good-natured eccentrics doing their own inimitable thing in isolation, then we have to break out of the ghetto, and take our place in the real world, and to do that we have to be taken seriously. The CFZ yearbook is our flagship publication, and I think that this year's volume is probably the best yet. It covers the whole range of CFZ activities across the globe, and although ever since the beginning we have included my annual report, in the 2009 volume, for the first time, are annual reports from the American and Australian offices, and a picture gallery of the year's events.

As far as the articles are concerned, it is heartwarming to see CFZ authors who have written for us ever since the beginning such as Dr Karl Shuker, Richard Muirhead, and Neil Arnold, alongside newcomers to the CFZ like Gary Opit and Theo Pajimans. Possibly the most contentious article herein is Michael Woodley's investigation of human evolution. So contentious is it, potentially at least, that I felt it necessary to write a special introduction for it explaining both Michael's and our motivations for writing and publishing it. Because, this is the sort of research to which I was alluding a few paragraphs ago; research which can easily be hijacked by people with dubious motivations.

The way that mainstream society seems to deal with such topics is to discourage research into them on the basis that such research could encourage racism. To us this is nothing short of scandalous. We already have a situation where various types of historical revisionism are now illegal. This is not the climate, nor the place, to discuss these matters of history, but the fact that genuine research can be - at best - frowned upon, and at worst - banned - in the name of political correctness, to us - at least -smacks of the ideological excesses of Nazi Germany, rather than the legitimate defence mechanisms of a civilised society. So we published it; on principle. Because, despite our many faults and weaknesses, the CFZ are men and women of principle, and we stand or fall by our moral beliefs.

But it is time to get off my soapbox. So, ladies and gentlemen, boys and girls, welcome to a new yearbook. I hope that you enjoy it and continue to support the CFZ for many years to come.

Best wishes

Jon Downes,
Director, CFZ
North Devon
January 6th 2009

STATEMENT OF CORE BELIEF

We believe:

1. That as Bernard Heuvelmans wrote "there are lost worlds everywhere" and that there are many new species of large animal awaiting discovery. Furthermore that new species of animal can be found even in the most highly urbanised environments.

2. That it is arrogant in the extreme for humankind to think that we know all the secrets of the universe, and especially those of the natural world. There are many processses of nature that we simply do not as yet understand.

3. That evolution is a living and dynamic process, and that as surely as new species of animals have evolved in the past, they are evolving in the present day, and that the process may happen faster than many people think. We therefore believe that many cryptids may be newly evolved species rather than prehistoric survivors.

4. That Young Earth Creationism, and the popular conception of Intelligent Design are fallacious. Although some of us are aethiests, others are deists, but those who believe in God believe in Creation NOT Creationism. We pledge ourselves to combat the disturbing rise in power of those who would teach superstition as scientific fact, and who are also linked to political movements that we believe to be dubious in the extreme.

5. That cryptozoology, and the other related subjects which we study are a true science and we shun the non-scientific belief systems of what is popularly called the paranormal, and in particular those which are quasi-spiritual in nature. Furthermore we carry out our researches with no religious or political bias and shun those who do. Any researcher who uses his/our research for such ends will be expelled from the CFZ forthwith.

6. That zooform phenomena and other allied phenomena are not `supernatural` or `paranormal` but are merely defined by laws of science that we do not yet understand. Because of this we are opposed to the present system of book marketing which include books on fortean zoology within the `Mind, Body and Spirit` category.

7. That our research is only justified if we publish our results within conventional as well as electronic formats, and make them freely available to everyone - not just those who are members of the CFZ.

8. That everyone involved with the CFZ became involved because of a childhood fascination with Natural History. For about a century from the mid 19th Century, Natural History was the most popular interest for British people of all ages and backgrounds. However, as a direct result of ill thought out government legislation since about 1970, this interest has decreased alarmingly. We feel that this is a bad thing and pledge to do all we can to redress this.

9. That the CFZ should be an international brother/sisterhood of like minded people who work together, mindless of differences of creed and culture, to push back the boundaries of human knowledge, for no other reason than that it is a good thing to do; hence our motto Pro Bona Causa Facimus (we do it for a good reason).

10. We believe that the study of fortean zoology is an ideal discipline for children and young adults, because it is an ideal introduction to the joy of increasing the sum total of human knowledge, because it is an unparalelled way of introducing the joys of the natural world to those who might not otherwise have experienced them, and because we believe that our modus operandi encapsulates many important intellectual disciplines sadly missing from much of modern

IN THE WAKE OF THE WINGED SEA SERPENT
Theo Paijmans

In his classic *In The Wake Of The Sea Serpents*, Bernard Heuvelmans proposed a classification into nine species of large marine animals – the so-called sea serpent – which he based on a study of 358 sightings.[1] Since his classification model first saw print it has been the subject to criticism [2], but apart from that, the oceans are a puzzling cosmos unto themselves. It is in the watery depths of the seemingly endless expanses of the seas that cover most of our small planet where the purely biological slowly merges into the fortean; where as on the maps of old there are monsters that defy even the most comprehensible methodology or classification model. And in the thousands of sightings of sea serpents we find a handful of accounts that may hint at the existence of an entirely new sub-species, that of the winged sea serpent. Truthful account, tall tale or sailor's yarn is something that only concerns us here insofar that it always serves as a warning, a let the reader beware, when digging up and studying as yet unknown cryptofortean accounts from the 19th and early 20th century newspapers. What these accounts perhaps may indicate, is that there actually may have been rare reports of even rarer sightings involving large marine creatures adorned with various unusual appendages described as 'wings'.

The first account that I was able to locate stems from a number of American newspapers that published between June and August 1881 an account of what a Captain Larsen allegedly saw:

> "Captain Larsen, of the bark *Honor*, which arrived here yesterday, reports that while about half-way between Madeira and St. Vegas, Canary Islands, he passed one of the most remarkable fish that he ever saw. This marine monster is described as being about forty feet in length, with four large fins, or wings, arranged in a row down its back.
>
> "These fins varied in length, according to the reckoning of Captain Larsen, from eighteen to twenty-two feet, and in width from six to nine feet. At the time of its being sighted the fish was about a quarter of a mile to windward of the vessel, and was lashing the water with its tail and wings, evidently in combat with some other monster." [3]

While the 19th century American newspapers published accounts of sea monster sightings in great numbers, even wryly commenting on 'the sea serpent season', we only find a hint of another remarkable occurrence published in a short notice in 1892: "A skipper claims to have seen the sea serpent in the sky. If the phenomenal monster has really been translated many an ancient mariner will experience a discouraging loss of material for new yarns." [4] But in the account of the capture of a strange marine animal that was published in 1896, there is some mention of a possible tradition among sailors that involves sea serpents with wings – that is, if we can take this report at face value:

> "Capt. George Belcher, while fishing near New Haven, Conn., the other day caught a remarkable sea monster. The animal is 4 feet 1 inch long, 3 feet 6 inches wide, and 4 inches thick. It is of a dark gray color, covered with hair, and has two wings. An old salt who saw it said: "Why that fish is nothing but a small sea serpent. It's got wings but hasn't fully developed yet." [5]

In 1901, when the coasting schooner *James Slater* arrived at Morris street wharf in Philadelphia with a cargo of bones and scrap iron from Havana, Edgar Hassann, steward of the ship, told a tale of 'weird experiences'. As the newspaper reported, included among these experiences was an encounter with a red sea serpent with wings, which followed the schooner for several miles:

> "The *Slater* left Cuba on the 11th of November with the largest cargo of bones ever shipped to this city... About a week after leaving Havana and while in the Gulf Stream Steward Hassann and George Peterson, the captain's brother, were startled to see a peculiar fish following the vessel a few yards behind. It was dark red in color, with its eyes on the side, large ears and had wings nearly a yard long on either side. The steward declares that he never saw or heard of such a monster, and the crew were greatly alarmed at its appearance. Several times it jumped out of the water and flapped its large fins, opening its mouth wide. The monster was no doubt attracted by the stench of the bones, and the crew feared that it would jump on board. When night came the strange animal suddenly disappeared and was not seen again.
>
> "The *Slater* is in command of Captain John James Peterson, who has been captain of the schooner for fourteen years. She carries a crew of ten men and is a three-master." [6]

THE HYDROPLANE SERPENT

We begin to discern something of the outline of the winged sea serpent. Not only are the sightings far and few between, but each time we encounter a new and different description. Either there must be in existence a whole subset of extremely large, winged marine creatures, or there are yarns and hoaxes, misidentifications and superstitions. Until 1911 I found no mention of the sighting of any sea monster with wings. That year:

> "Passengers and the crew of the White Star liner *Celtic* brought with them to New York today a revival of the sea serpent tales of other years.

THE WONDERFUL STORY TOLD BY CAPT. LARSEN.

A HOLE KNOCKED IN HIS VESSEL BY A HUGE SOMETHING WITH TAIL AND FINS—THE TERRIFIED SAILORS ESCAPE IN BOATS AND ARE PICKED UP BY A STEAM-SHIP.

The steam-ship P. Caland, which arrived from Rotterdam yesterday, had on board Capt. L. C. Larsen and the entire crew of a Norwegian bark which was lost in mid-ocean under remarkable circumstances. Capt. Larsen's story, which is fully corroborated by his men, goes to show that the accounts which have been given to the public from time to time of dangerous sea monsters inhabiting the ocean are not wholly false. Either the original sea serpent or some powerful submarine creature of the same race, ran into the bark Columbia on Sept. 4, in latitude 47° 32', and longitude 43° 54', and injured that vessel so that she sank within a short time after she had been struck. The bark Columbia left London for Quebec, in ballast, on Aug. 8, and met with variable weather until the morning of the 4th inst., when she was sailing at the rate of from six to seven miles an hour before a fair wind. The sea was not very rough, and the bark was carrying all sail and heading westward. The Captain was on deck at five bells in the morning watch, when a tremendous shock, which shook the bark from stem to stern, was suddenly felt. The men

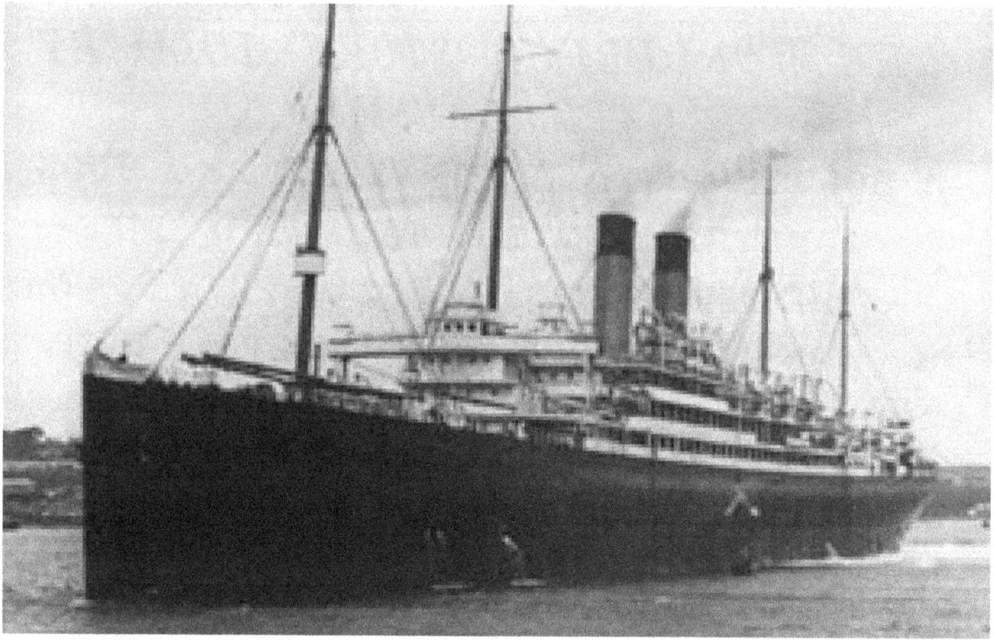

"R. M. S. CELTIC".
20904 Tons, 697½ feet long, 75½ feet broad, 49 feet deep.

"They reported having passed early yesterday morning a formidable looking creature which was going at high speed in pursuit of a school of young whales. The monster, they say, had wings, and rose frequently 10 feet or more from the water. Whales and pursuer faded from sight within a few minutes." [7] The improbable tale prompted one of the many American newspapers that carried the story to state with its tongue lodged in its cheek that "The sea serpent season has arrived and the first reports tell of a winged monster that chases whales. Unless the winged monster is larger than an aviation field it will be in order to ask what it will do with the whales it may succeed in capturing."[8], and another dryly observed that "Even the sea serpents appear to recognize the necessity of keeping right up in the line of progress." [9]

Fun notwithstanding, there seems to have been a second ship involved in the sighting of this most unusual winged sea serpent, as a number of newspapers reported.

"What are the strange marine creatures passengers on ocean steamers see, that are loosely described as sea serpents? The sailor men on the same vessels either shortly answer "Whales!" or decline altogether to give any response to the inquirer. A few days ago a steamer came into New York several of whose passengers reported that they had distinctly observed the strangest of odd fishes, a monstrous creature that had wings and generally resembled a partially submerged biplane. It was menacing of aspect and inspired those onlookers who were landsmen with a creepy feeling. Stewards of the steamer interviewed by reporters gave the stereotyped explanation "Whales at play!"

"Now comes a story from Philadelphia which cannot have been concocted by passengers of two vessels without such expenditure for wireless telegraphy as few makers of fiction would care or dare to incur. While the New York story is vague, but contains a sketch or hint of a creature with biplane attachment, the observers on board steamer *Haverford* just arrived at Philadelphia describe the monster of the deep as "a hydroplane serpent."

"When appealed to, Captain Thornton of the *Haverford* does not deny the soft impeachment that he saw a monster fish, but refrained from reporting the occurrence lest he incur the ridicule every mention of the sea serpent evokes from the incredulous who have never been on salt water. Captain Thornton seems to have no gift for the clear presentation of detail, but some of the passengers are almost graphic in their narration. They say the monster had "a gray body mottled with black and purple, and was nearly as long as the Haverford." Now here is something approaching a statistical basis for approximating the characteristic of this weird denizen of the deep. The *Haverford* is a steamer belonging to the Morgan combine, under the British flag, a regular ocean liner, Lloyds Register gives the Haverford's gross tonnage as in excess of 11,000 tons, and her extreme length as 531 feet. Hence if the monster, fish or serpent was nearly as long as the Haverford, the extreme length of said monster came very close to being one tenth of a mile. What was it? Whales do not attain such proportions, even in the versions of the old whalers... Great sunfishes, inert, looking like an expanse of gray blanket reposing on the surface, are occasionally encountered at sea, but the biggest of these are estimated to weigh considerably less than a ton a piece..." [10]

Five years later, in 1916, the month being August, the officers of the Wilson Line steamship *Colorado* on its way from Hull to Boston, saw a very strange sea creature at the time the liner was crossing the Grand Banks. There,

"...the attention of the chief engineer was attracted to what appeared to be the sails of a small fishing schooner. Suddenly the supposed sails flapped down with great force against the surface of the water. Then they were raised and brought down again. This was repeated several times. The chief engineer called Capt. Collins.

"What appeared to be sails were the fins of a sea monster, whose huge curving back could be made out occasionally in the wash of the sea.

"Capt. Collins said that the strange looking fish was apparently battling with a whale, which it had seized by the back and was lashing with its great fins. Both Capt. Collins and the chief engineer are old seafaring men and they declared yesterday that they had never seen anything like the attacking monster.

"Several whales were sighted during the passage and a fight was witnessed between a trasher and a whale. The trasher seemed to be getting the best of the tussle, for it would lift its huge tail out of water and bring it down with a resounding smack on the broad back of the whale. The latter was trying to escape, but the trasher pursued it and sank its teeth repeatedly into the whale's side.

"Despite fog the Colorado made the passage from the North Sea port in less than 12 days, one of the fastest on record." [11]

Interestingly, another newspaper described the monster as looking "...like the sails of a small schooner" and "a gigantic fish with... far-reaching wings..."[12]

List of passengers for the *Haverford* August 10, 1910, voyage.
Captain Thornton in command.

WINGED SEA SERPENT REPORTED IN LEVANT

Currant-Laden Felucca Sunk When Sailor Belts Reptile With Pomegranate.

HAS EYES LIKE SIDELIGHTS

Zooms Aloft Like Airplane—Wrecks Caciques With Figs and Pistachio Nuts.

After a long absence, the sea serpent has appeared once more. According to reports just received from Constantinople and Smyrna, the elusive reptile is disporting itself in the placid waters of the Mediterranean, terrifying fishermen and crews of the feluccas and other small craft which ply along the shores of the Levant.

The most thrilling story was brought here last week by officers of the National Greek liner Constantinople, who said that motor launches armed with one-pounder quick-firing guns were searching the Sea of Marmora for a winged marine monster which had been sighted circling Dog Island at great speed for three or four days.

When it rose from the sea and flew over San Stefano toward the city of Stamboul it was reported to have made a booming sound like the German triplanes that flew over Paris during the war. How the huge marine monster passed through the Dardanelles without causing a big wash along the shore has not yet been discovered.

Officers on the Constantinople said that the flying sea serpent was first reported off the ancient port of Chalcis in the Aegean Sea and so alarmed the sponge divers that an appeal was sent to Athens to send a gunboat to search for the monster and destroy it. When the commander of the fort fired a gun in the direction indicated by the scared sponge divers the serpent rose to an altitude of about 5,000 feet and flew away toward the port of Smyrna, Asia Minor, where it was reported next day to have sunk two caciques laden with figs and pistachio nuts.

Zubdee Effendi, a Turkish rug merchant who arrived from Smyrna on the liner, said that he saw the scaly-winged monster quite plainly as he was crossing the bay to his home on one of the islands. When the head rose from the surface of the water, he said, all the natives on board flung themselves on the deck and prayed, while the vessel rocked wildly and the sea was covered with white foam as if it was being violently agitated underneath.

The head of the nautical reptile was fully ten feet across, with two enormous reddish green eyes butting out on either bow like a ship's sidelights, while its immense flappers looked as if they easily weighed a ton each. Its beam across the middle of the back, which rose high out of the water, Zubdee Effendi continued, was fifteen feet and the length over all was fully fifty feet.

When the sea serpent turned its head toward the boat the heat on deck became unbearable, said the Turkish rug merchant. He felt that his beard was shriveling up. Finally the monster dived, flung its huge tail into the air with a swish that nearly sent the small schooner over on its beam ends and then disappeared from view.

Within a few days the flying sea serpent, Zubdee Effendi said, was reported off the mouth of the Nile at Damietta, the port of Candia in Crete, and off the island of Mitilene, where it was stated to have rammed and sunk a felucca laden with currants because one of the crew struck the reptile in the eye with a pomegranate. Sponge divers and pearl fishers in the Aegean Sea are reported to be taking a rest until the sea serpent has been destroyed.

THE SCALY WINGED SEA MONSTER

We will have to wait another six years, for a new and even more incredible report to emerge in the press. In 1922 several American newspapers reported on the sighting of what was termed a 'winged marine monster'. And while the first sentence of the reports is not true – I have in my files a multitude of sightings of the more or less expected and ordinary sea serpent sightings in a continuous succession between 1916 and 1922 – the rest of the account makes one scratch ones head as to the veracity of the account.

> "After a long absence, the sea serpent has appeared once more. According to reports just received from Constantinople and Smyrna, the elusive reptile is disporting itself in the placid waters of the Mediterranean, terrifying fishermen and crews of the feluccas and other small craft which ply along the shores of the Levant.
>
> "The most thrilling story was brought here last week by officers of the national Greek liner *Constantinople*, who said that motor launches armed with onepounder, wuick-firing guns were searching the sea of Marmora for a winged marine monster which had been sighted circling Dog island at great speed for three or foru days.
>
> "When it rose from the sea and flew over San Stefano toward the city of Stamboul it was reported to have made a booming sound like the German triplanes that flew over Paris during the war. How the huge marine monster passed through the Dardanelles without causing a big wash along the shore has not yet been discovered.
>
> "Officers on the *Constaninople* said that the flying sea serpent was first reported off the ancient port of Chalcis in the Aegean sea and so alarmed the sponge divers that an appeal was sent to Athens to send a gunboat to search for the monster and destroy it. When the commander of the fort fired a gun in the direction indicated by the scared sponge divers the serpent rose to an altitude of about 5000 feet and flew away toward the port of Smyrna, Asia Minor, where it was reported next day to have sunk two caciques laden with figs and pistachio nuts.
>
> "Zubdee Effendi, a Turkish rug merchant who arrived from Smyrna on the liner, said that he saw the scaly-winged monster quite plainly as he was crossing the bay to his home on one of the islands. When the head rose from the surface of the water, he said, all the natives on board flung themselves on the deck and prayed, while the vessel rocked wildly and the sea was covered with white foam as if it was being violently agitated underneath.
>
> "The head of the nautical reptile was fully ten feet across; with two enormous reddish green eyes butting out on either bow like a ship's sidelights, while its immense flappers looked as if they easily weighed a ton each. Its beam across the middle of the back, which rose high out of the water, Zubdee Effendi continued, was 15 feet and the length over all was fully 50 feet.
>
> "When the sea serpent turned its head toward the boat the heat on deck became unbearable, said the Turkish rug merchant. He felt that his beard was shrivelling up. Finally the monster dived, flung its huge tail into the air with a swish that nearly sent the small schooner over on its beam ends and then disappeared from view.

"Within a few days the flying sea serpent, Zubdee Effendi said, was reported off the mouth of the Nile at Damietta, the port of Candia, in Crete, and off the island of Mitilene, where it was stated to have rammed and sunk a felucca laden with currants because one of the crew struck the reptile in the eye with a pomegranate. Sponge divers and pearl fishers in the Aegean sea are reported to be taking a rest until the sea serpent has been destroyed." [13]

As is so often the case with these puzzling cryptofortean accounts, atles and reports, I found no further mention of the creature and any action that the local authorities might have undertaken, considering the slight possibility that this story is more than the wild tale of a rug merchant, and we can't even vouch for his existence.

We have read how in 1896 a strange creature of the deep was netted and caught, and we find a similar event having taken place in 1929. In that year apparently, fisherman Thomas Bowen and four others caught a very odd specimen after a long and arduous struggle:

"Beach Haven, N.J. - A fish ceases to be a fish when it grows wings spreading nine feet, develops eyes in its ears, a five-foot tail and two feet each equipped with two toes. It becomes a sea monster. According to the tale by Thomas Bowen, fisherman, it was a sea monster and not a fish that caught in his net. Finding the wight too great for him, Bowen called for four other fishermen to help him. The five struggled from shortly before seven in the morning until eleven. The monster they finally beached weighed 440 pounds, measured nearly twelve feet and presented a cream-colored front and a royal purple back, the story goes." [14]

Now if only some newspaper would have published a photo of this, judging from the description, incredible denizen of the deep. Perhaps such a photo does exist in a local newspaper that is outside the scope of what I am able to survey. I located the last account of a winged sea serpent in a few 1938 newspapers, shortly before the outbreak of the second world war. Short and cryptic as these reports oftentimes were published, as a mere filler in between other items on a newspaper page, we may wonder what it was that the startled crew of the fishing boat saw off Gloucester: "Boston, July 28 – (INS) – Sea serpents have gone streamlined and modern. The crew of the fishing boat Giuseppe today avowed that thirty-five miles off Gloucester they saw a sea monster of the 1938 model. It was 50 feet in length, had a head like a horse, and a set of wings like an airplane." [15]

And with this account we have reached the last of the handful of alleged reports of encounters with a very unusual family of aquatic beings. Based on these few accounts with descriptions that differ in each case, one may feel inclined to conclude that this set of anomalous reports of dubious value leave very little that can be considered as any evidence.

On the other hand we find in these fragmentary tales and reports for the existence of a winged sea serpent a certain motif, that of the transportation of the idea of the legendary winged dragon from land to the sea.

Moreover, when studying the lake monster reports that also were published in great numbers in the 19th and 20th century American newspapers with great regularity, we encounter, embedded in this set of tales, accounts of winged lake serpents and in even somewhat larger numbers than their saltwater counterparts. Possibly the case for the winged sea serpent truly and only belongs to modernday folklore, to the traditions, legends and yarns of the seas. However, what we do have in front of us is a largely forgotten chapter in the complex saga of the great sea serpent and how it was perceived, handed over and written down, and perhaps in this we find that it ultimately serves as an echo or reminder that, as these stories try to convey, in real life the seas and their inhabitants represent a far stranger universe than we can ever imagine. And that we, the land creatures, may be able to build large vessels forged of iron and steel, but in our vanity we have by no means conquered nor even begun to comprehend that mighty otherworld, the oceans, at all.

Notes

1. Bernard Heuvelmans, *In The Wake Of The Sea Serpents*, Hill and Wang, 1968, pages 537 – 572.
2. Ulrich Magin and Lars Thomas, 'St George Without A Dragon', Steve Moore eds., *Fortean Studies Vol. 3*, John Brown, 1996, pages 223-236.
3. 'A Sea Monster', *Daily Inter Ocean*, Chicago, Illinois, 18 June 1881; *Alton Telegraph*, Alton, Illinois, 4 August 1881; Ohio, Athens, 4 August 1881; *The Athens Messenger*, Athens, Ohio, 4 August 1881; *The Indiana Progress*, Indiana, Pennsylvania, 18 August 1881.
4. *The Daily Inter Ocean*, Chicago, Illinois, 26 August 1892.
5. 'Sea Serpent Season', *Fort Wayne Gazette*, Fort Wayne, Indiana, 7 June 1896.
6. 'Schooner Was Owned By Bugs. Strange Marine Monster Also Encountered by the Slater on Trip From Havana', *Philadelphia Inquirer*, Philadelphia, Pennsylvania, 12 December 1901.
7. 'Sea Serpent With Wings. Marine Monster, Seen by Celtic's Passengers, Chases Whales ', *The Washington Post*, Washington, District Of Columbia, 5 June 1911; 'They Were All Sober But Saw Sea Serpent. The Monster Had Wings And Frequently Cleared The Water Ten Feet. An All This Is True, Say Passenger and Crew of Steamer Celtic, Arriving at New York Yesterday Morning', *Miami Herald*, Miami, Florida, 5 June 1911; 'Winged-Serpent After Whale. Passengers And Crew Of White Star Liner Tell Of Terrible Monster Seen At Sea', *The Montgomery Advertiser*, Montgomery, Alabama, 5 June 1911; 'Tales Of Sea Serpents Again Brought Here', *Duluth News-Tribune*, Duluth, Minnesota, 6 June 1911; 'Winged Sea Serpent Sighted At Sea. Celebrated Monster of the Deep Seen By Passengers of Good Ship Celtic; Was Chasing Whales at High Speed', *Albuquerque Journal*, Albuquerque, New Mexico, 5 June 1911; 'Sea Serpent Appears. Passengers of Celtic Report Sighting Formidable Looking Creature', *Trenton Evening Times*, Trenton, New Jersey, 7 June 1911; 'Flying Sea Serpent', *The Evening Telegram*, Elyria, Ohio, 25 August 1911.
8. *The Waukesha Freeman*, Waukesha, Wisconsin, 8 June 1911.
9. *Oelwein Daily Register*, Oelwein, Iowa, 18 July 1911.
10. 'Wonders Of The Deep. Passengers on Ocean Steamers See Strange Marine Creatures Supposed to be Sea Serpents', *The Newport Daily News*, Newport, Rhode Island, 14 June 1911.
11. 'A Strange Monster. Seen From Liner – Fins First Appeared To Be Small Sails', *Lowell Sun*, Lowell, Massachusetts, 29 August 1916.
12. 'Weird Sea Fight Story In Port. Steamship Captain Tells of Strange Sea Monster That Spanked a Whale', *Boston Journal*, Boston, Massachusetts, 29 August 1916.
13. 'Winged Sea Serpent Reported In Levant. Currant-Laden Felucca Sunk When Sailor Belts Reptile With Pomegranate. Has Eyes Like Sidelights. Zooms Aloft Like Airplane – Wrecks Caciques With Figs and Pistachio Nuts', *New York Times*, New York, New York, 13 August 1922; 'Sea Serpent with Wings Reported', *Syracuse Herald*, Syracuse, New York, 15 September 1922; 'Winged Sea serpent Reported In Levant. Felucca Sunk When Sailor Pelts Reptile With Pomegranate', *New Castle News*, New Castle, Pennsylvania, 19 September 1922; 'Sea Serpent With Wings', *Appleton Post-Crescent*, Appleton, Wisconsin, 6 October 1922; 'Flying Sea Serpent Once More Comes To

Cause Consternation. Officers of Greek Liner Tell Thrilling Story of Monster Hunted With Rapid-fire Guns by Men in Motorboats', *Oregonian*, Oregon, 22 October 1922.
14. 'Fisherman Tell of Landing Sea Dragon', *The Van Wert Daily Bulletin*, Van Wert, Ohio, 8 October 1929; 'Fisherman Tell of Landing Sea Dragon', *The Kingston Daily Freeman*, Kingston, New York, 24 October 1929.
15. 'Sea Monsters Go Streamline', *The Charleroi Mail*, Charleroi, Pennsylvania, 28 July 1938; '1938 Sea Serpents', *Lowell Sun*, Lowell, Massachusetts, 28 July 1938; 'Sea Serpent, 1938 Model, Reported', *San Antonio Light*, San Antonio, Texas, 28 July 1938.

STAR ROT: DEBRIS FROM A FALLING STAR? OR JUST A LOAD OF ROT?

Jonathan Downes

In November 2006, my colleagues and I at the Centre for Fortean Zoology were in the pub in the tiny North Devon village in which we live and work. We found ourselves talking to one of the local farmers. He had come up to us, diffidently asking whether we were "those blokes who hunt monsters"? When we replied that we were, he sat down with us, and over a pint or three we talked about some of the strange fortean zoological phenomena which have occurred in North Devon over the years.

The English countryside, or at least our particular corner of it, is quite a strange place. There have been so many sightings of the mysterious black cat, that there is hardly a soul in the village who would argue against its existence. Frogs of a bright golden hue are seen hopping in and out of the vicarage fishpond, and within the village there are so many ghost stories that my long-suffering fiancée is determined to collect them all together in book form.

I don't know quite how the conversation got on to the subject of star rot. It is an obscure phenomenon, and one which has not attracted much interest in recent times. In the 17th Century, the philosopher and poet Henry More wrote the following lines:

"the Starres eat those falling Starres, as some call them, which are found on the earth in the form of a trembling gelly, are their excrement".

And as recently as 1846, *Scientific American*, an unashamedly fuddy-duddy journal, which is the oldest continually published magazine in the United States, and one of the most respected of its type in the world described a similar occurrence:

"It appeared larger than the sun, illumined the hemisphere nearly as light as day. [And when it fell] a large company of the citizens immediately repaired to the spot and found a body of fetid jelly, four feet in diameter,"

The phenomenon has always interested us because, despite what is claimed within current scientific thinking, anomalous objects - and even living creatures - *do* fall from the heavens. The phenomenon is often referred to by the unlovely and cumbersome acronym of fafrotskies [fish and frogs raining out of the skies]. This happens more often than many people are prepared to admit.

It has often been suggested that the phenomenon of star rot is somehow linked with these mysterious creature falls. Such things have continued to intrigue us because, despite the best efforts of scientists and lay-men alike, nobody has ever come up with a convincing explanation for them. The most popular explanation is that the fish and frogs have been sucked into the upper atmosphere by waterspouts, which then proceed to dump them upon unsuspecting eye-witnesses. This doesn't hold water – if you will excuse the pun – for a moment. As my friend, Dr. Mike Dash, pointed out in his excellent book *Borderlands: The Ultimate Exploration of the Unknown* (Heinemann, 1997) fish and frogs are not the only things found in ponds. There is a singular lack of reports of rusty shopping trolleys, muddy Wellington boots, car tyres, rotting cigarette packets, traffic cones or abandoned fridges falling upon the heads of unsuspecting passers-by.

The conversation then meandered on to other things, as conversations in pubs often do, and eventually we supped-up, and went our respective ways.

A few days later, we received a 'phone call from our farmer friend. "Guess what?" he said. "You know that ol' star rot you were talking about? Well I have got some in one of my fields."

Sadly, I had a previous engagement, and so, while I was engaged in a tedious discussion with my solicitor, it was Richard Freeman and Mark North, the Zoological and Art Directors of the Centre for Fortean Zoology, who went in my stead.

We are a non-profit making organization, and so we cannot afford unlimited numbers of vehicles. In this case, it would not have made any difference, because neither Mark nor Richard can drive. Luckily, Mr. Heywood, the farmer who had reported the case of star rot to us, was passing through the village with an empty sheep truck, so our intrepid investigators, emulating one of my favourite scenes from *An American Werewolf in London*, hopped aboard. It was the day before Hallowe'en, and winter was only just around the corner. Although one can't really complain about the weather in Devon, a windy October day is not the most pleasant of times to be bumping up country lanes and cart tracks in the back of a rattly old sheep truck. It was drizzling, and the irony is that by the time our heroes arrived at Duerdon Farm, they were equally as disheveled as Richard had been after several weeks of trekking through the jungles of Sumatra in search of the fabled Orang Pendek.

They parked in the farmyard, and walked, with Mr. Heywood, across two or three fields towards the isolated area where the star rot had been reported. As they walked, they discussed some of the other explanations that have been mooted over the years to explain this peculiar phenomenon. One enduring explanation is that it is nothing more or less than frogspawn, laid in fields where a pond used to exist by frogs who have the geographical location hard-wired into their brains. Frogs do, indeed, do this. For many years there was a large concrete pond at the bottom of the garden here at the CFZ, but when my brother and his wife had small children, my late father removed the pond, fearing that while his grandchildren were at such an inquisitive age, if he did not do so tragedy was bound to ensue.

For some years after the pond was removed, frogs – certainly from the lineage which had spawned and bred in the pond for nearly thirty years – continued to lay their eggs on the lawn. Only for them to wither and die.

However, to the best of my knowledge there have been no reports of foetal nuclei and in any case, they hypothesise even with global warming having upset the natural rhythm of the seasons, frogs do not lay their eggs as early as October. No. They would have to look elsewhere for an explanation.

Another theory is that star rot is nothing more than the partially digested body of a jellyfish that had been eaten and then regurgitated by a passing sea bird. Again, this is highly unlikely. Jellyfish are insubstantial creatures at the best of times, and are not normally the food items for birds. So that's another theory out of the window.

The belief in star rot is particularly prevalent in Wales. The Welsh peasants have always believed that when a meteor falls to the earth it becomes reduced to a mass of jelly. They call this jelly *pwdre ser*. It is also known as "star-slough," "star-shoot," "star-gelly" or "jelly," "star-fall'n" and it appears in farmers' fields, most often in the autumn; the season when meteor showers are most readily observed. As early as 1641 Sir John Suckling (1609-1642) wrote the following lines, which well describe the way in which these gelatinous substances came to be regarded as the remains of a "fallen star":

'As he whose quicker eye doth trace
A false star shot to a mark'd place
Do's run apace,
And, thinking it to catch,
A jelly up do snatch.'

As Richard was reciting this poem, they finally reached the field, and there, in the middle of the dank autumn grass, were a number of lumps of what did – for all the world – look like semi-solidified frothy saliva. Richard and Mark took samples, photographed and filmed the strange crusty patches, conducted a brief interview with Mr. Heywood, and retraced their steps back to the farmyard.
They walked back to the village, discussing the matter as they did, and as soon as they got back to the office they telephoned Gordon Rutter. It was my mother who, in many ways, set me on the path I wished to follow in my life. She gave me my love of books, my love of animals, and somewhat of a peculiar taste in literature for someone of my generation. At the age of seven, completely bored with the adventures of *Janet and John*, my mother introduced me to the pulp fiction of the late 19th and early 20th Centuries. Amongst the books to which she introduced me, were the novels of Leslie Charteris featuring a rakish hero called The Saint. In one of these books, called 'Boodle' (1934), the hero claims that whereas some men collect stamps, and others collect first editions, his hobby was collecting strange friends. Even at the age of seven, I was impressed and set out to do likewise.

Gordon Rutter is someone I have known for many years. He also happens to be one of the world authorities on the folklore and the science of fungi. He immediately identified the photographs as *Fuligo septica* – a species of slime mould.

Slime moulds are peculiar organisms called protists – uncategorisables that are neither animal, plant nor fungi. Even now, some zoologists consider them to be animals and some mycologists still consider them to be fungi. They are little understood, but what we do know about them seems to mirror something from the darkest episodes of *Dr. Who*. Fuligo is part of a group called plasmodial slime moulds – they usually take the form of single cell amoebi but under certain conditions can mass together into the 'feeding stage' which is what Richard and Mark had encountered on Duerdon Farm. This is a giant amoeba with thousands of nuclei, not divided by cell membranes, but enclosed by a single outer one. They can move at a rate of about one millimeter an hour, scavenging to engulf their food, which includes bacteria, fungi, yeast and decaying organic matter.

> Charles Fort had several bizarre and gloriously imaginative theories to explain such strange falls from the heavens. His most imaginative idea was the concept of what he called the 'Super-Sargasso Sea'. Just as the Sargasso Sea in the North Atlantic is supposed to be full of shipwrecks and all manner of objects caught up in its gulfweed, so the 'Super-Sargasso Sea' might be a repository for terrestrial and extra-terrestrial matter high above the earth's surface. Sometimes the sea would suck things up; at other times it would spew them back down to earth. He also formulated the theory of 'teleportation' - a force capable of transporting objects and animals from place to place without traversing the intervening distance. This has been used to explain several anomalies including creature-falls and the anomalous appearance of various out-of-place animals which appear in places that logic and the accepted dicta of conventional zoology suggests that they should never be.

At least, for the moment, the mystery was solved. Richard and Mark had gone in search of fafrotskies, and come back with samples of one of the most peculiar, and - to this author at least - sinister, organisms known on the planet. We have tried to grow it under laboratory conditions, but failed spectacularly. However, this has whetted our appetite, and next autumn we shall be scouring the farmer's fields once again.

Because, although on one level the mystery has been solved, in another sense it is only just beginning. Although we have identified the cause of one particular outbreak of star rot, we have not solved the mystery as a whole. We know, from conversations with other local farmers, that the phenomenon is quite well spread.

However, it would seem that the slime mould we retrieved from Mr Heywood's farm may not be responsible for the occurrences in other farms in the area; these seem to be much more glutinous and jelly-like. Is there a second species of slime mould which enters its feeding stage in conjuction with the meteorite showers of the 'season of mists and mellow fruitfulness'? And what is it that actually triggers these slimy amalgamations of otherwise single-celled organisms to infest farmers' fields? And do they really coincide with meteorite showers?

Mr Heywood's star rot did; it appeared right in the middle of the Orionid meteor shower. And could it be that the reason we were unable to cultivate the slime mould under laboratory conditions was that the meteor shower was over? Like a pop song of my youth said: there are more questions than answers, and like Lao Tzu – a Chinese philosopher from the 6th Century BC – once wrote: 'The more we see the less we know'.

Watch this space.

> A few months before Richard and Mark's adventure, *New Scientist* magazine announced that researchers from the University of Southampton and the University of Kobe in Japan have even used slime mould to control the behaviour of a bizarre six-legged robot. They grew the plasmodium in a six-pointed star shape on top of a circuit and connected it remotely via a computer to the robot. Any light shone on sensors mounted on top of the robot were used to control lights shone on to the mould.
>
> As the slime mould tried to get away from the light, the robot scrabbled away into the darkest corners. This level of complex behaviour is something that nobody had ever considered in such a primitive organism.

THE TARASQUE
Richard Muirhead

The Tarasque is a semi-mythical creature thought to have lived from between nearly two thousand years ago up until about the Middle Ages in the lower Rhône Valley of France. It is one of those legendary hybrid "monsters" whose existence "resonates" far beyond its home, said to be in the marshy Camargue region of France to be precise, with the town of Tarascon being the main focal point of the procession of the Tarasque model. (See Figure 1 below, a postcard in this author's possession showing the Tarasque parade in the early years of the 20th Century.) By "resonates" I mean there are representations of the Tarasque beyond the confines of France's Camargue district, e.g. in English churches. In a different branch of human knowledge, the Tarasque has become a bone of contention in the debate between creationists and supporters of the theory of evolution.

Fig.1

The word 'Tarasque' derives from the verb *tarir* "to drain, to dry up" [1], though the author I am quoting from, T.Tindall Wildridge, writing this towards the end of the nineteenth century, did not give a reason why this draining aspect was associated with the creature. Could it be that its movement through the waters displaced or "drained" the river around it? This would suggest that the Tarasque was a heavy bodied creature. The basic outline of the Tarasque legend is as follows:

Provençal Christian tradition recounts that Martha, Marie-Magdalene, Lazarus and other saints (as well as fourteen bishops) were thrown into an open boat with no sails or oars in A.D. 50 Palestine. They were said to have landed in the Camargue around 48A.D. Martha went up the Rhône and reached Tarascon where a dreadful monster lived in a den beside the river. On being implored by the people of Tarascon, St.Martha captured the monster and handed it over to them. They tore it to pieces and then were converted to the new religion. St Martha settled in Tarascon, where she died in 68A.D. The relics of St Martha were hidden to avoid the destructive rage of the Saracens. In 1199, a beautiful Romanesque church was consecrated in her name; the very fine porch still remains. Many fervent pilgrims came to pray before the tomb of the Saint and it was thus quite natural that monasteries, the homes of prayers and meditation, grew up in this area. [2]

This creature 'had to be an amphibious animal of great size, perhaps a crocodile.' [3] The possibility that the Tarasque was a crocodile is surely a contender amongst a range of candidates. Crocodiles may have survived in parts of Europe until the 19th Century. For example, according to Trottman, referring to research in Italy, 'stuffed crocodiles and dragon bones can be found in the so called cabinets of curiosities. They are usually located in churches or other ecclesiastical buildings.' [4]

Earlier Trottman had remarked that in Ulrich Magin's "recent publication" *Die Seeschlange vom Comer See* this latter cryptozoologist refers to 'various crocodile sightings and legends in the rivers and lakes of northern Italy.' [5]

Much earlier than Magin, Druce has studied the crocodile in the European medieval ecclesiastical environment and he indicates that knowledge of the crocodile can be shown from church architecture in Yaxley, Hunts, Topsham, Devon, Bury St. Edmunds and Kilpeck, Herefordshire. [6]

Also, according to two letters in *Country Life* on March 17th and May 12th 1977 respectively, from Mary Huddleston, there is (or was) a motif in a church in Hadfield, Derbyshire resembling a Tarasque and there was folklore connected with the image. It showed a half dragon, half beast with a wing folded on its back to its side. Ben Chapman, a misericord expert, could not identify it and cast doubt on the veracity of this story. [7]

Ettinger had further insights into the Tarasque legend as far as its echo in Britain is concerned. She has pointed out that a Romanesque capital dating from c.1180 in York and an English relief 'on the font in the Church of St John the Baptist, at Stone, in Buckinghamshire' [8] date unknown, can both be traced back to the legend of the Tarasque. This legend came to Britain from Liguria in Italy as a form of the heroic Hercules, fighting an amphibious and terrestrial monster. Like the Tarasque.

Returning to the continent, a visitor to Brno in the Czech Republic (as it is now) in 1849 saw the body of 'an alleged lindworm, or dragon, preserved from a very remote period' and hanging in an arched passage leading to the town hall. Describing this dragon in Notes and Queries, the author said he had been unable to examine the creature closely, but affirmed 'it is undoubtedly of the crocodile or alligator species.' [9]

In France, it has been stated that 'the skin of the dragon slain by St Bertrand (in the 12th Century) was preserved in the cathedral at St Bertrand de Comminges in Haute Garonne, and really belongs to a small crocodile or alligator species.' [10] A contributor to Chambers Book of Days mentions that 'in churches at Marseilles, Lyons, Ragusa and Cimiers (in France) skins of stuffed alligators are exhibited as the remains of dragons'; [11] Unfortunately the author doesn't say whether all those animals are supposed to have been killed in Europe.

De Smet has observed:

'It is possible that crocodiles used to live along the coast in river deltas or swamps of France, Spain and the Balearic islands until the first centuries of our era. Lavauden (1926) supposes that the 'Tarasque', a mythical monster of the Rhône delta, really was a Nile crocodile, living in the Camargue' [12]

Crocodiles lived in the Sahara up to at least 1858 [13] 'In Roman times, the Nile delta abounded in crocodile.' [14] The crocodile was present until at least 1912 in what is now northern Israel, for example near the foot hills of southern Mt. Carmel. [15] Crocodiles occurred in Sicily from time to time: 'The chronicles of Sicily mention that crocodiles sometimes appeared on the island, in the rivers Papireto and Garaffello near Palermo and in the Amenano near Catania'. [16]

In Spain, giants accompanied by a dragon or an eagle, in the Corpus Christi processions of Spain are thought to derive from 'the underwater monster or tarasca.' [17] Returning to the United Kingom: Miss K. MacDonald saw in February 1932 an 8 foot long creature splashing up the River Ness. The description was remarkably similar to a crocodile. Adrian Shine thinks that this "crocodile" was in fact a sturgeon. [18] This is referred to by Andreas Trottman and Rupert Gould.

Louis Dumont, in his penetrating study *La Tarasque* (1951), sums up the reasons for [the Tarasque's] popularity under two main aspects: it is a symbol and focus of the town's sense of its identity and unity; and its mock-aggressive assaults on the crowds offer a healthy release to destructive instincts, transforming them into a vigorous, beneficient and luck-bringing energy.

Everything that we know of the English hobby-dragon, 'skirmishing among the crowd', 'the delight and terror of the children' and of adults who 'both feared him and loved him', shows that what is true of La Tarasque must also have been true of him. [19]

The description and etymology, etc of the Tarasque, to be found in Eberhart's *Mystery Creatures Vol 2 N-Z* [20] differs sharply from the opinions expressed above that the Tarasque may have been a crocodile. The etymology of 'Tarasque' is: 'From the castle of Tarascon, on the Rhône River'. Alternatively, Tarascon (originally called Nerluc) is said to have taken its name from the Dragon after it was killed. [21] That is straightforward. However, the description of the creature demolishes the crocodile thesis: (According to Dr. Darren Naish, a cryptozoologist and expert in dinosaurs, crocodiles died out in Europe millions of years ago. *) Eberhart goes on to give another description of the Tarasque: 'Size of an ox. Head like a lion's. Ears like a horse's. Hard skin, covered with spikes. Six legs (emphasis my own.) Bearlike claws. Serpentine or scorpion-like tail.' [22] Eberhart gives the following information:

Behaviour: Amphibious. Sloughs its skin every 7 years. Said to have caused the river to flood. Made itself a nuisance by eating people and destroying bridges. [23]
Habitat: an underwater cave near Tarascon. [24]
Distribution: The Rhône River, between Arles and Avignon, Provence, France. The animal is said to have come originally from Galicia in central Turkey, which may indicate a Celtic origin. [25]
Significant sightings: St Martha (a Syrian prophetess conflated with Martha, the sister of Lazarus) was said to have overcome Tarasque with holy water and the sign of the cross. There were reports of river

monsters in the Rhône in 1954 and 1955. [26]
Present status: The city celebrates St Martha's victory over Tarasque with a festival in late June each year.
Possible explanations:

1. A Nile crocodile (*Crocodylus niloticus*), especially since St Martha is associated with the Middle East.
2. An auroch (*Bos primigenius*), though this wild European bull was neither amphibious nor particularly ferocious.
3. Creationists have suggested that Tarasque was the Late Cretaceous dinosaur Triceratops, though the legend does not mention horns on the head. A closer match would be a glyptodont, a large armadillo-like mammal that lived in South America until the end of the Pleistocene, 10,000 years ago. [27]

There are many iconographical representations of Martha and the Tarasque in churches across Europe e.g. at Lignieres- the Tarasque in chains. [28] This author (that is, Richard Muirhead) sympathises with the dinosaur survival model, though he recognises, at the same time, that no one dinosaur fits into the description of Tarasque satisfactorily. Bill Cooper's comments on *Beowulf*, an ancient Nordic poem, could easily apply to the Tarasque: If instead of Beowulf we insert the word "Tarasque" and use French instead of Danish, then we will gain new insight into this legendary French creature:

It is clear from all the evidence that we are dealing in *Beowulf* with an account that is entirely and unambiguously historical. The poem treats of men and animals that have a firm place in history, with events that are real, and which occurred in real places. It is also clear that we are dealing here with accounts of creatures that the Danes should have known nothing of if what the evolutionists say is true, that dinosaurs and man did not co-exist. Thanks to the *Beowulf* epic and several other ancient sources, we know that they surely did, and may yet co-exist today. [29]

Any comments on this essay can be made to Richard Muirhead at richmuirhead@tiscali.co.uk Special thanks to Maggie Hughes for translations from the French, Pierre-Yves Le Pogam at the Musèe du Louvre Paris and the staff of Macclesfield Library for help with providing material.

REFERENCES

1. T.Tindall Wildridge. *Animals of the Church in wood, stone and bronze.* (Loughborough: Heart of Albion Press,1991),p.10
2. Official web site of Tarascon www.tarascon.org quoted in e-mail from Andreas Trottman to Richard Muirhead August 1st 2008.
3. Trottman Ibid.
4. E-mail from A.Trottman to Richard Muirhead August 19th 2008
5. E-mail from A.Trottman to Richard Muirhead August 18th 2008
6. G.C.Druce *The Symbolism of the Crocodile in the Middle Ages* pub in *Archaeological Journal* vol.66. 1909 Plate IV facing p.316
7. R.Muirhead notes from *Country Life* March 17th 1977 and May 12th 1977
8. E.Ettinger *The Ligurian Hercules and La Tarasque Ogam* vol.16 1964 p158
9. `R.S Jun` N & Q series 1 vol 2 p.517 in M.Behrend *The Dragon as Crocodile Geomancer* p.17
10. "A.H.S." Book review. *Folklore* vol 9 p.73 in Behrand Ibid pp 17-18
11. Anon article on April 23rd in *Chambers Book of Days* vol 1 pp 540-1 in Behrend Ibid p.18

12. L.Lavauden *Les vertèbrès du Sahara*. (Tunis: Albert Guènard,1926) in K.de Smet *Status of the Nile crocodile in the Sahara desert* Hydrobiologia 391 1999 p84
13. K.de Smet Ibid p.83
14. K.de Smet Ibid p.83
15. K.de Smet Ibid p.84
16. J.Anderson *Zoology of Egypt vol 1 Reptilia and Batrachia* (London: Bernard Quaritch,1898) in K.de Smet Ibid p.84
17. R.Partington *The Annual Feast of Santiago da Compostella*. In *Folklore* vol. 68(2) June 1957 p.361
18. E-mail from A.Trottman to Richard Muirhead August 19th 2008 citing R.T.Gould *The Loch Ness Monster.*
19. J.Simpson *British Dragons* (Wordworth Editions in association with The Folklore Society,2001),p.117
20. G.M.Eberhart *Mystery Creatures vol.2 N-Z* (Santa Barbara: ABC-Clio Inc, 2002),p.536
21. G.M.Eberhart Ibid p.536
22. G.M.Eberhart Ibid p.536
23. G.M.Eberhart Ibid p.536
24. G.M.Eberhart Ibid p.536
25. G.M.Eberhart Ibid p.536
26. G.M.Eberhart Ibid p.536
27. G.M.Eberhart Ibid p.536
28. L.Rèau. *Iconographie de L`Art Chrètien Tome III Iconographie des Saints* II G-O (Paris: Presses Univertaires De France,1958),p.894
29. B.Cooper *Beowulf The Monsters and the Man.* (Portsmouth: Creation Science Movement,2008),p.4

* Conversation with Dr Darren Naish August 2008

PROLOGUE TO MICHAEL WOODLEY'S ARTICLE

In the 14 years I have been editing CFZ Yearbook, I don't think that any article has caused so much discussion amongst the CFZ leadership. It is not because they article is particularly scientifically contentious, but because of the political, and socio-political implications that it has. Indeed, according to some interpretations of current human rights legislation, this article may even be illegal. If so, it is quite possibly the most stunning indictment of a politically correct society that we have ever seen.

The 1964 UNESCO declaration on Race and Racial prejudice reads, in part:

1. All men living today belong to a single species, Homo sapiens, and are derived from a common stock. There are differences of opinion regarding how and when different human groups diverged from this common stock.

2. Biological differences between human beings are due to differences in hereditary constitution and to the influence of the environment on this genetic potential. In most cases, those differences are due to the interaction of these two sets of factors.

3. There is great genetic diversity within all human populations. Pure races-in the sense of genetically homogeneous populations-do not exist in the human species.

4. There are obvious physical differences between populations living in different geographical areas of the world, in their average appearance. Many of these differences have a genetic component. Most often the latter consist in differences in the frequency of the same hereditary characters.

5. Different classifications of mankind into major stocks, and of those into more restricted categories (races, which are groups of populations, or single populations) have been proposed on the basis of

hereditary physical traits. Nearly all classifications recognise at least three major stocks.

One can see how this statement, coming as it did, less than two decades after the horrific events of Auschwitz and Belsen, made perfect political sense. Here, I would like to state categorically, once and for all, that from a socio-political point of view, neither I, nor anybody else involved in the CFZ, can find anything wrong with the political thinking behind this statement: all people on this planet are indeed equal in the rights of the law, and this is one of the few unquestionably human rights. However, I personally have always been interested in the research which suggests that from a *scientific* point of view the 1964 declaration is quite simply untrue. Like many creditors zoologists, I believe that we are not alone on this planet.

I believe that we are not the only extant member of the genus Homo, and I believe that within my lifetime, at least two other species of human - the almasty, and the strange beings who inhabit some of the swamps in Texas and the other southern states - will be proven, not only to exist, but to be our closest relatives.

When this happens, I want to see them receive the full protection of national, and international, human rights legislation. The way things are at the moment, with all existing human rights legislation based around the 1964 declaration, our putative cousins will receive no protection at all under human rights legislation, and will be treated like animals.

This is both wrong, and counter to the spirit of the 1964 declaration. I, like so many from Jesus downwards, believe in upholding the spirit of the law, rather than the letter of the law, and hope that documents like Michael Woodley's analysis of possible human speciation should be seen in a positive light, rather than as any attempt to reinforce negative racial stereotyping.

Because, the human race - whether one, or more species - is/are particularly unpleasant. Research like Michael Woodley's has often, in the past, been hijacked by the marching morons of the racist right-wing, who try to use it to justify their illogical and unpleasant ideology. I would be appalled, as would Michael Woodley, if anything of the sort happened as a result of our publishing his research paper.

Jonathan Downes
3rd January 2008

Far across the ocean
In the land of look and see
There once was a time
For you and me

Where the winds blow sweetly
And the easy seas flow still
And where the barefoot dream of life
Can laugh and cry its fill

Where slot machine confusions
And the plastic universe
Are objects of amusement
In the fiction of their curse

And where the crazy whiteman
And his teargas happiness
Lies dead and long since buried
By his own fantastic mess

For I hate the whiteman
And his plastic excuse
For I hate the whiteman
And the man who turned him loose…

I hate the White Man
      Roy Harper

HUMAN DIVERSITY AND HOMINOLOGY
Michael A. Woodley

Introduction

Novel primate taxa continue to be discovered and described frequently in the modern era. Examples include the Highland mangabey *Rungwecebus kipunji* (Jones *et al.*, 2005) and the GoldenPalace.com monkey *Callicebus aureipalatii* (Wallace *et al.*, 2006). These, and other recent discoveries indicate that the primate description record has not reached maximum enrichment and that there are therefore reasonable grounds for speculating on the existence of as yet undiscovered and undescribed primate taxa.

A number of ethnoknown primate species are currently objects of cryptozoological study; including the Sasquatch, Yeti, Alma and the Yeren. These cryptid primates - studied collectively within the branch of cryptozoology known as hominology, are of special anthropological interest, as they appear to possess human morphological affinities, indicating a possible close evolutionary relationship.

A variety of speculative identity theories have been advanced in an effort to taxonomically situate these cryptids. Krantz allied the Sasquatch with the Gigantopithicines (Krantz, 1986), speculating that they probably crossed the Bering land-bridge from Asia into North America at around the same time as modern humans (Krantz, 1992).

Meldrum's studies of the pressure patterns present in casts made by Sasquatch indicate the possibility that they possess an ape rather than human arrangement of foot bones. This has lead to speculation that the evolution of bipedalism in apes need not be accompanied by major rearrangements of the bones of the foot as is currently believed, and that this particular trait may represent an adaptation to bipedalism unique to human lineages (Meldrum, 2004).

A group of cryptid primates described by Heuvelmans as the 'Neanderthaloid Wildmen', which include the Siberian Alma and the Chinese Yeren, have been situated within the genus *Homo* by a number of authors. Based on his examination of the remains, Heuvelmans suggested that the famed Minnesota Iceman, the specimen of which was allegedly obtained in Indochina during the Vietnam War, represented the holotype of a new *Homo* species possessing Neanderthal affinities – *Homo pongoides* (Heuvelmans, 1969). Heuvelmans subsequently demoted the Wildman to the status of subspecies - *Homo neanderthalensis pongoides*, after collaboration with Porshnev (Heuvelmans and Porshnev, 1974). Porshnev had previously suggested that the Wildmen be designated as *Troglodytes recens* (Porshnev, 1963; 1969).

In an article published in *Current Anthropology*, Porshnev declared his objective to rewrite the postulates of primatology and anthropology through his theory that modern humans had one immediate ancestor in the form of *Paleanthropus* (*Troglodytes fossilis*), representatives of which survive into the modern era as the Wildmen (Porshnev, 1974). Perhaps unsurprisingly, anthropologists were largely unsympathetic to Porshnev's calls for such radical and apparently unwarranted revision (Simpson, 1984).

It has additionally been speculated that legends of the dwarfish Sumatran Orang-Pendek may have a more substantial basis in fact owing to the recent discovery of *Homo floresiensis*, an extinct hominid of short stature, which lived on the Indonesian island of Flores until at least 18,000 years ago (Gee, 2004).

In their search for novel hominoid and hominid species hominologists investigate the most extreme wilderness environments in an effort to locate what must be very rare and elusive organisms, if indeed they even exist; the thesis of this manuscript, in contrast to traditional work in hominology, attempts to address the admittedly controversial but scientifically legitimate question of whether 'hidden' (unrecognized) taxa may exist as part of the extant natural diversity present within the taxon that we choose to call *Homo sapiens sapiens*.

In addressing this question, recent molecular and phenetic findings will be considered, coupled with an investigation into whether taxonomic schemes are consistently applied when considering the true extant diversity within this taxon.

Models of human origin I: multiregionalism.

There are two major scientific models of human origins. The older of the two models is known as the multiregional hypothesis.

This hypothesis holds that there was a single migration event approximately two million years ago involving *Homo erectus* leaving Africa and coming to colonize most of Eurasia. The hypothesis suggests that these populations were connected via gene flow, which resulted in disparate *H.erectus* populations eventually evolving into *H.sapiens* (by way of the Neanderthals in the case of European populations) (Coon, 1962).

The hypothesis also suggests that; traits generally useful to humans such as high intelligence and the capacity for language spread rapidly throughout the full clinal range of these hominids, in part because such traits would have facilitated gene flow by permitting greater contact between separated populations (Thorne and Wolpoff, 2003).

Not all traits would have been universally beneficial however as some such as disease resistance and skin colour would have conferred adaptations to conditions encountered locally, which would have facilitated the evolution of racial level variation (Coon, 1962; Thorne and Wolpoff, 2003).

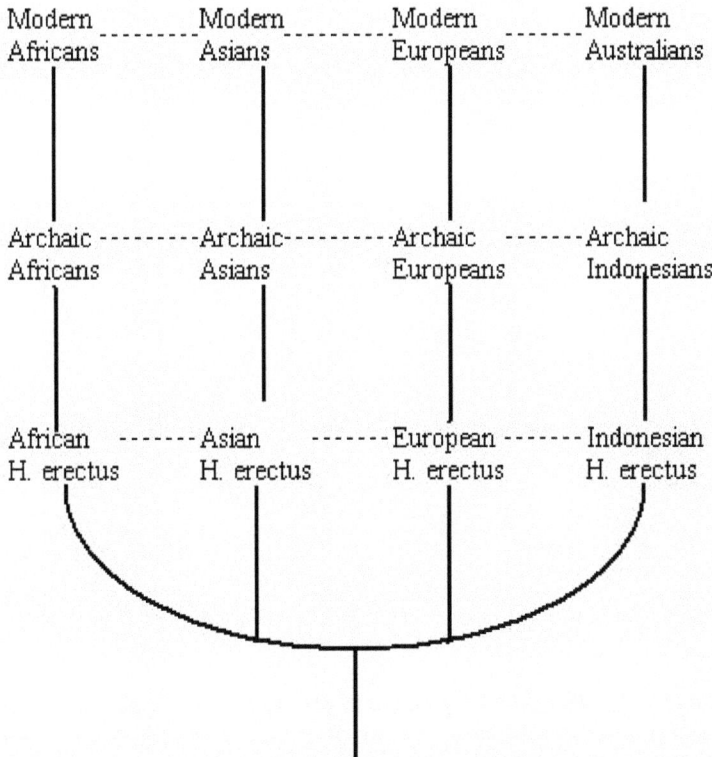

Figure 1: A diagrammatic illustration of multiregional evolution in humans. The initial H. erectus *migration event is represented at the base of the tree where it splits. Each population of* H. erectus *can be seen to be evolving into the four classical extant major racial groups of humanity (top of the tree). The dotted lines illustrate gene flow between the various populations at various stages of evolution (Image retrieved from* http://en.wikipedia.org/wiki/Image:Multiregionaltheory.gif).

The multiregional hypothesis remained the dominant theory on human origins until the early 1990's, when it was challenged by the out-of-Africa hypothesis, which posited a single, relatively recent African origin for *H.sapiens*.

Models of human origins II: out-of-Africa.

The out-of-Africa hypothesis (known also as the recent single-origin hypothesis, or the replacement hypothesis) is based on the assumption that humans had a single origin point in Africa, and that after evolving from *H.erecuts* between 200,000 and 100,000 years ago, various groups broke off and migrated from Africa into Europe and Asia 60,000 years ago, where they effectively came to replace established populations of Neanderthals in Europe and relict *H.erectus* in Asia (Stringer and Andrews, 1988). There is some debate as to the exact mechanism of replacement. It has been suggested that a likely cause was direct competition between the different taxa, which may have resulted in one species simply successfully out competing the others through a combination of being able to better utilize existing resources (developing superior hunting tools, more elaborate social networks etc) and the use of warfare, which resulted from modern humans simply being more aggressive than their competitors – a hypothesis based on the so called killer ape theory first advanced by Dart in the 50's (Dart, 1953). Another, more recent theory suggests that hybridization may have occurred between Cro-Magnon (very early European *H. sapiens*) and Neanderthals, both of which coexisted approximately 40,000 years ago (Trinkaus and Shipman, 1983), although there is much debate over how extensive this may have been if indeed it even occurred (Krings *et al.*, 1997).

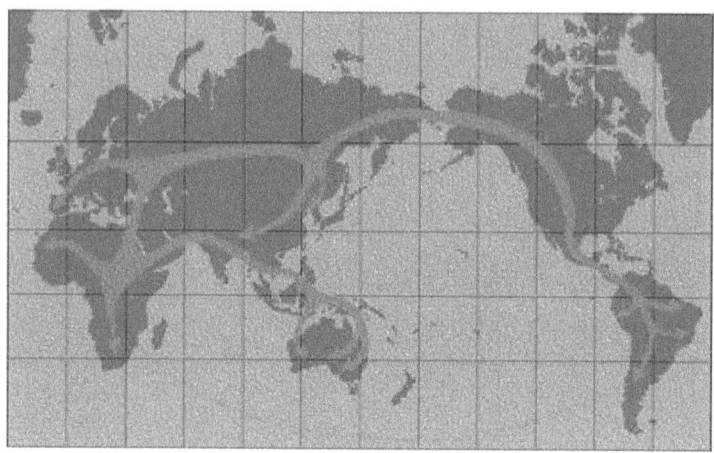

Figure 2: A Mercator map illustrating the path taken by H. sapiens *60,000 years ago as it left Africa and began to spread around the world (Image retrieved from http://www.progonos.com/furuti/MapProj/ Dither/ProjAppl/projAppl.html).*

The out-of-Africa hypothesis has, as was previously mentioned, come to largely supplant the multiregional hypothesis as the dominant model of human origins. Evidence for a relatively recent African origin for *H.sapiens* is compelling. For example studies that involve tracing the ancestry of mitochondrial DNA (mtDNA - which is inherited matrilinealy) suggest that every person may be related to a single female ancestor, termed 'mitochondrial Eve', who lived in Africa as recently as 140,000 years ago (Cann *et al.*, 1987). Despite this however there also exist compelling data, which seem supportive of multiregionalism, such as the discovery of an ancient Australian aboriginal skeleton known as 'Mungo Man', which despite being anatomically modern and at most 40,000 years old, possesses mtDNA from a lineage with no contemporary descendants. Advocates of multiregionalism have suggested that mtDNA does not in fact reflect ancestry or divergence time, as the genes are themselves subject to natural selection (rather than being selection neutral). If validated, these findings would directly challenge both the out-of-Africa and the mitochondrial Eve hypotheses (Adcock *et al.*, 2000).

The race debate.

The term 'race' has historically been used in biology as a synonym for 'subspecies' or 'variety'. Whereas the terms 'subspecies' and 'variety' have often been used in the description of infra-specific variation in non-human animal and plant lineages respectively, the term 'race' tends to be employed exclusively in the description of the same levels of variation present within the human species. The four major definitions of what constitutes a subspecies or race have been reviewed by Long and Kittles (2003).

Concept	Definition	Author(s)
Essentialist	"A great division of mankind, characterized as a group by the sharing of a certain combination of features, which have been derived from their common descent, and constitute a vague physical background, usually more or less obscured by individual variations, and realized best in a composite picture."	Hooton (1926)
Taxonomic	"An aggregate of phenotypically similar populations of a species, inhabiting a geographic subdivision of the range of a species, and differing taxonomically from other populations of the species."	Mayr (1969)
Population	"Races are genetically distinct Mendelian populations. They are neither individuals nor particular genotypes, they consist of individuals who differ genetically among themselves."	Dobzhansky (1970)
Lineage	"A subspecies (race) is a distinct evolutionary lineage within a species. This definition requires that a subspecies be genetically differentiated due to barriers to genetic exchange that have persisted for long periods of time; that is, the subspecies must have historical continuity in addition to current genetic differentiation."	Templeton (1998)

Table 1: The four major definitions of race as reviewed by Long and Kittles (2003).

Table 1 illustrates the evolution of classificatory concepts of race from essentialist to lineage based. Although in each case the idea of 'distinctness' is invoked as a necessary criterion for the existence of a race there exists considerable disagreement over how to define and operationalize (make quantitative) that distinctness. The essentialist concept of Hooton places the emphasis on the existence of combinations of characteristics shared through common descent, whereas the taxonomic concept uses a combination of phenotypic similarity coupled with the idea of range restriction. The population concept of Dobzhansky on the other hand talks of race exclusively in terms of Mendelian populations whilst the lineage concept of Templeton requires races to have been subject to gene flow barriers whilst simultaneously exhibiting historical continuity.

i) Critical race theory.

The table would seem to suggest that there is no universally agreed upon definition of race or subspecies and that the use of any particular race concept in the apportionment of human biological diversity is to a degree arbitrary. This suggestion of arbitrariness has led many social scientists to claim that what is termed 'race' is in fact nothing more than a 'social-construct', devoid of any biological foundation. According to this view, which is known generally as critical race theory, the concept of racial classification is a recent invention (c. 18th century) and was designed as a means of grouping subjugated colonial peoples on the basis of arbitrary physical characteristics. By this logic the very notion of race therefore has inherently racist connotations as, it is inferred, the decision to use concepts of race in the 'arbitrary' grouping of humans is suggestive of a desire to delineate an out-group that is some way 'inferior' in contradistinction to a 'superior' in-group to which, it is presumed, the classifier would belong (Delgado and Stefancic, 2001).

The problem with critical race theory is that it attempts to engage racial classification on a normative rather than a scientific level. The suggestion that scientific race concepts such as those reviewed in table 1 stemmed from a desire on the part of their respective

Footnote 1: As evidence of the pervasiveness of this view within the social sciences, a 1985 survey of 1,200 academics who were asked whether they disagreed with the statement: "There are biological races in the species *Homo sapiens*", revealed that only 16% of biologists disagreed as compared to 53% of socio-cultural anthropologists (Lieberman *et al.*, 1992). The likelihood is that an even higher percentage of social scientists would disagree today. As evidence of this, one only needs to read the official position statements on race and ethnicity of major organizations such as the American Anthropological Association and the American Sociological Association.

authors to apportion people into 'inferior' vs. 'superior' categories is simply an appeal to motive and on those grounds alone such suggestions can be simply ignored as fallacious from a scientific perspective. The notion of arbitrariness in the taxonomic classification of races is a significant and legitimate scientific issue in need of redress however, one which in fact has significant implications for the taxonomy of *H.sapiens*.

ii) Construct validity and operationalization.

Prior to examining the race concept from a classificatory stand point it is necessary to demonstrate its validity as a biological construct independently of classificatory schemes. It was mentioned previously that all four of the major race concepts require races to be in some way distinct from one another, in this section this distinctness will be operationalized and it will be shown that the four major race concepts are in fact ultimately compatible with one another.

It is frequently asserted that because the majority of genetic variation (85%) lies within the classically defined racial groups rather than between them (15%), race is therefore a taxonomically meaningless category. Lewontin, who first articulated this idea, essentially assumed that because there is a 30% probability of misclassifying an individual's race based on the variation in a single genetic locus, race must therefore be categorically invalid (Lewontin, 1973). Edwards has however countered this suggestion with the observation that although Lewontin's arguments are correct for a single locus, the conclusions to which he came were essentially fallacious as the likelihood of misidentification rapidly approaches 0% as more loci are considered. This is due to the fact that loci frequencies within racial groups tend to be correlated (Edwards, 2003).

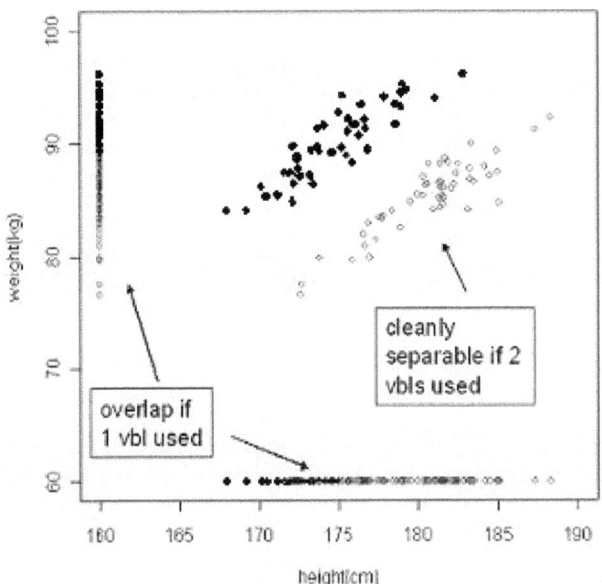

Figure 3: A graphic illustration of the so-called 'Lewontin's fallacy'. In this example there are two hypothetical races (black and white points) defined by different combinations and distributions of weight and height. If information about only a single variable (analogous to a genetic locus) is used to identify a race then the identity is partially obscured as a result of overlap. If information about two variables is considered however then the races emerge as cleanly separable and distinct clusters (Image retrieved and modified from http://www.gnxp.com).

Based on figure 3, race and synonymous concepts can be defined basically as a composite number of traits whose distributions intercorrelate in such a way so as to give rise to a particular, distinct correlative structure. This basic definition allows for a potential reconciliation of the four major attempts at defining race listed in table 1. Hooton's essentialist definition, which requires the sharing of characteristics through common descent is clearly compatible with the observation that race is a correlation structure of traits, as is Mayr's taxonomic definition, which sees races as phenotypically similar groups occupying different ranges. Eco-geographical distinctions between races would be to a degree congruent with respect to genetic and phenotypic traits, so would be expected to yield correlation structures similar to figure 3. There is no reason why the correlation structures could not correspond to Mendelian populations as is required by the population definition of Dobzhansky, nor is there any reason why the distinct correlation structures could not have been subject to restrictions in gene flow, or exhibit some degree of historical continuity as is required by the lineage definition of Templeton. These last two would in point of fact be a prerequisite for the evolution of racial differences in the first instance. The four major race concepts can therefore be united within a common operationalizational framework, the differences between them are purely a matter of where the qualitative descriptive emphasis is placed.

iii) Comparative classification.

Demonstrating the biological construct validity of race does not necessarily address the issue of classification.

Although it has been shown that the four major attempts at defining race differ only in terms of qualitative descriptive emphasis, the problem of taxonomic arbitrariness in terms of how diversity within species is apportioned and classified is still an issue. To illustrate the inconsistency with which race and synonymous concepts have been used in the apportionment of infra-specific diversity, comparative measures of genetic diversity for a range of species along with the numbers of recognized extant subspecies are presented in the following table.

Racial variation within human populations is frequently dismissed as existing only at infra-subspecific scales (if indeed it is even acknowledged to exist at all) and *H.sapiens* is typically described as monotypic (comprising one species and one subspecies).

Based on table 2, it is evident that this view is inconsistent with the way in which taxonomic classification has been employed for other species exhibiting similar degrees of heterozygosity.

Chimpanzees for example exhibit very similar degrees of observed heterozygosity to humans (0.63-0.73 vs. 0.588-0.807) yet have been divided into four subspecies. Some species such as the grey wolf actually exhibit lower levels of observed heterozygosity than humans (0.528 vs. 0.588-0.807) yet have been divided into as many as 37 subspecies.

Based on a superficial interpretation of these data *H.sapiens* should be expected to contain anywhere between three and 37 subspecies, possibly more, however *H.sapiens* is unique, at least amongst primates in that 85% of its total diversity exists within racial groups rather than between them. In the cases of most animals exhibiting a subspecies population structure the majority of the diversity exists between the subspecies.

This would seem to indicate that estimates of the number of subspecies within *H.sapiens* should be expected to be on the low side.

In order to conservatively estimate the true number of subspecies within *H.sapiens*, the four traditionally recognized great races (Africanoid, Caucasoid, Mongoloid and Austroloid) could be used as a ball park figure, the accuracy of which is largely corroborated through the molecular analysis of classical and other genetic markers, which consistently reveal the presence of around five major clades in the form of Sub-Saharan Africans, Caucasians, Greater Asians, Australopapuans and Amerindians (Nei and Roychoudhury, 1993).

Species (vernacular name)	H_e	H_o	Number of recognized extant subspecies	Author(s)
Humans	-	0.776	1	Wise et al (1997)
Humans	-	0.7-0.76	1	Jorde et al (1997)
Humans	-	0.588-0.807	1	Bowcock et al (1994)
Chimpanzees	0.78	0.73	4	Reinartz et al (2000)
Chimpanzees	-	0.63	4	Wise et al (1997)
Bonobos	0.59	0.48	1	Reinartz et al (2000)
African buffalo	0.759	0.729	5	Van Hooft et al (2000)
Leopards	0.36-0.8	-	Between 8 and 18 depending on the preferred taxonomy	Uphyrkina et al (2001)
Jaguars	0.739	-	9	Eizirik et al (2001)
Pumas	-	0.52	6	Culver et al (2000)
Canadian lynx	-	0.66	3	Schwartz et al (2002)
Polar bears	0.68	-	1	Paetkau et al (1999)
Brown bears (N. America)	0.26-0.76	0.3-0.79	19	Paetkau et al (1998)
Brown bears (Scandinavia)	0.709	0.665	19	Waits et al (2000)
Coyote	0.675	0.583	19	Garcia-Moreno et al (1996)
Gray wolf (N. America)	0.62	0.528	37	Garcia-Moreno et al (1996)
Dogs (42 breeds)	0.616	0.401	1	Garcia-Moreno et al (1996)
African wild dogs	0.643	-	5	Girman et al (2001)
Dingo	0.47	0.42	1	Wilton et al (1999)
Wolverines (N. America)	0.42-0.68	-	Between 2 and 3 depending on the preferred taxonomy	Kyle and Strobeck (2001)
Wolverines (Scandinavia)	-	0.27-0.38	Between 2 and 3 depending on the preferred taxonomy	Walker et al (2001)
Elk (N. America)	0.26-0.53	-	Between 7 and 8 depending on the preferred taxonomy	Polziehn et al (2000)
Bighorn sheep	0.681	0.566	3	Forbes et al (1995)

Table 2: Comparative values of genetic diversity for a variety of mammalian species representatively sampled across their respective ranges (except where indicated), as measured by autosomal microsatellites ([H_e] = expected heterozygosity; [H_o] = observed heterozygosity). Table modified from Goodrum (2002).

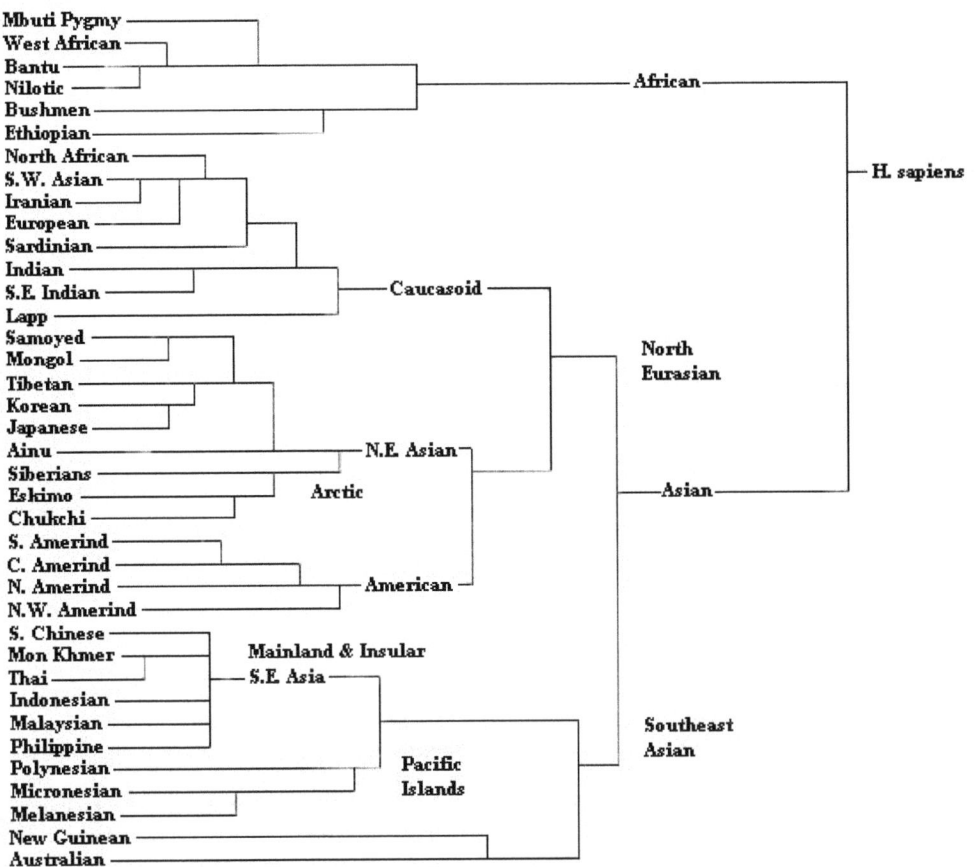

Figure 4: The human phylogeographic tree illustrating the relationships between 38 ethnic groups divided between five major clades (continental populations); African, Caucasoid, N.E. Asian, American and Australian/New Guinean. (Cavalli-Sforza, 1991).

Based on the above observations, a compelling case could be made for these five major clades qualifying as subspecies through a consistent application of taxonomy.

Even by the classical standards of phenetics, racial morphological distinctions appear to be well defined. Sarich and Miele have noted that these differences are on average about equal to the distances among species within other genera of mammals, with the exception of populations generated under domestication pressures, such as breeds of dog. They also note that they are unaware of any other mammalian species where the constituent races are as strongly marked as in *H.sapiens* (Sarich and Miele, 2004).

In a sense, rather than clarifying the issue of classification as it pertains to race, it has been demonstrated instead that it is arbitrary for no other reason than the inconsistency with which the subspecies concept is evidently employed. Clearly, within the context of classical taxonomy, it is better to discuss the apportionment *and* operationalization of racial diversity in terms of major clades correspondent with subspecies and derived through the use of genetic and/or phenotypic markers, which show high degrees of consistency irrespective of which markers are used. The use of a cladistic definition of subspecies not only compliments the definition of race as correlation structure, but also presents a potential solution to the problem of arbitrariness in traditional taxonomic approaches to the classification of human racial diversity.

Are there multiple extant human species?

In answering this question it is necessary to ignore, for the moment, the possible existence of relict Neanderthal or *H.erectus* populations and focus instead on the extant diversity within *H.sapiens*. A minority of anthropologists in the past have held the view that human racial differences are great enough in some instances to warrant being considered as species level differences (*see eg.* Cartwright, 1857); however these views were often based upon the use of scientifically inappropriate morphological comparisons with extant primates (such as degree of prognathism). In this section, the two major definitions of species will be considered in addressing this question.

i) Species concepts.

As with the concept of race, there have been multiple attempts made at defining 'species'. There are around 13 different species concepts which, like with the four major race concepts, can mostly be shown to differ primarily in terms of where the descriptive emphasis is placed. As far as vertebrate classification is concerned, there are currently two major classification paradigms: the traditional biological species concept and the phylogenetic species concept, which also happen to be amongst the most diametrically different of the species concepts.

Table 3 illustrates the differences between the biological species concept of Mayr, which regards species as the end products of an evolutionary chain of events that have lead to the establishment of reproductively isolated populations; and the phylogenetic species concept, first introduced by Eldridge and Cracraft, which sees species as being defined in terms of the evolutionary distinctiveness of lineages. There have been attempts to redefine phylogenetic species. Eldridge and Cracraft (1980) saw synapomorphic characteristics (shared characteristics derived from a common ancestor) as the unit that both united and defined the smallest aggregate population or lineage, in other words the phylogenetic species. Mishler and Theriot (2000) have however suggested that the phylogenetic species is in fact that least inclusive taxon in a formal phylogenetic classification. The various phylogenetic species concepts can ultimately be shown to be highly similar, differing only in terms of descriptive emphasis, they also explicitly reject the existence of subspecies as a valid level of classification.

ii) Races as phylogenetic species.

A valid question to ask is what are human races in terms of the phylogenetic species concept? An argument could be made that they do not count as true phylogenetic species owing to the fact that admixture is occurring between the races, thus their evolutionary distinctiveness is being diminished. However, this can be countered with the observation that admixture between racial groups is a generally rare phenomenon, even when racial groups are living in close proximity to one another, and that the likelihood of admixture has been observed to be a function of the degree of genetic similarity between the racial groups involved (Morton *et al.,* 1967; Rushton, 2005).

As phylogenetic species represent the least inclusive monophyletic taxonomic unit within classical taxonomic schemes the concept dispenses with hierarchical classification altogether. Based on this classification, there exist only phylogenetic species grouped based on shared synapomorphic characteristics. Ow-

Concept	Definition	Author(s)
Phylogenetic	Species are the result of clear divergence within a group of organisms sharing an ancestor whose lineage remains intact with respect to other lineages throughout time and space. Subspecies are not recognized.	Eldridge and Cracraft (1980)
Biological	Species are comprised of populations that either have the potential to or actually interbreed, and are reproductively isolated from other such populations.	Mayr (1969)

Table 3: The biological and phylogenetic species concepts.

Species (Vernacular names)	Species (Scientific names)	Genetic distance	DNA type	Author(s)
Sub Saharan African (Bantu) vs. Australopapuan (Aborigine)	H.sapiens	0.33	Single Nucleotide Polymorphism (SNP)	Salter (2003)
Sub Saharan African (Bantu) vs. Eurasian (English)	H.sapiens	0.24	SNP	Salter (2003)
Human vs. Neanderthal	H.sapiens, H.sapiens neanderthalensis	<0.08	mtDNA	Caramelli et al (2003); Curnoe and Thorne (2003); Gutiérrez et al (2003)
Human vs. H.erectus	H.sapiens, H.erectus	0.17	mtDNA (inferred)	Curnoe and Thorne (2003)
Western gorilla vs. Eastern gorilla	Gorilla gorilla, G.bereingei	0.02-0.29	mtDNA	Jensen-Seaman (2000); Guillén et al (2005)
Weatern gorilla vs. Eastern gorilla	Gorilla gorilla, G.bereingei	0.38	14 nuclear loci	Thalmann et al (2006)
Common chimpanzee vs. Bonobo	Pan troglodytes, P.paniscus	0.05-0.2	mtDNA	Guillén et al (2005)
Common chimpanzee vs. Bonobo	Pan troglodytes, P.paniscus	0.49-0.68	Autosomal DNA	Becquet et al (2007)

Note: There is some debate as to whether or not the western and eastern gorillas constitute different species (see Thalmann et al., 2006).

Table 4: Comparative pair-wise genetic distances (expressed in terms of variance – F_{ST}) between various biological species (and subspecies) as listed in Fuerle (2008) and other sources.

ing to the fact that the majority of the genetic diversity within *H.sapiens* exists within classically defined racial groups rather than between them, the true diversity of phylogenetic species can be expected to be high. If the terminal groups (minor clades) of Cavalli-Sforza's phylogeographic tree (see figure 4) are used as the least inclusive taxonomic grouping, then there can be expected to be as many as 38 extant phylogenetic species comprising humanity.

iii) Are there unrecognized biological species within *Homo*?

To the knowledge of this author only a single attempt has been made at explicitly addressing this issue in recent years, namely the attempt of Fuerle (2008). Fuerle uses comparative genetic distance data involving various DNA types obtained from a variety of sources for a range of biological species and subspecies. The results of his review are summarized in the following table. Additional data involving non-mtDNA based estimates of the genetic distance between the gorilla species and the chimpanzees and bonobos have been included for comparison.

An extremely superficial interpretation of table 4, would suggest that the Sub Saharan African (Bantu) and Australopapuan (Aborigine) genetic difference as measured by SNP's is greater than the genetic distance between both the two species of gorilla (*Gorilla gorilla* and *G.beringei*), and greater than the distance between the common chimpanzee and the bonobo as measured by mtDNA.

On the basis of this Fuerle suggests that there are only two consistent courses of action to take regarding re-classification – splitting or lumping. Either *H.sapiens* could be split into two species – *H.africanus* which would encompass modern African populations and *H.eurasianensis* which would encompass Eurasian populations; making the genus *Homo* consistent in his view, species-wise with respect to other genera in which the differences between species are expressed in terms of much smaller genetic distances; or alternatively the genetic variability within the human species could be used to typologically define the absolute limits of what constitutes a vertebrate species, which could then be employed as a taxonomic baseline in the classification of other species. This would mean lumping the two gorilla species and the chimpanzee and the bonobo as single species.

The issue of lumping hominids has been taken seriously by Cunroe and Thorne (who produced the genetic distance estimates for the Neanderthals and *H.erectus* used in table 4). Echoing Porshnev's (1974) suggestion for a radical classificatory transformation of primatology and anthropology, they suggest that chimpanzees be reclassified into the genus *Homo,* and that living humans and all fossilised homos be reclassified as *H.sapiens*. Based on this they suggest that modern *H.sapiens* therefore diverged from the last common ancestor around 1.7 Ma. Their findings are indicative of the possibility that humans did not go through a recent genetic bottleneck, as per the mitochondrial Eve hypothesis, and that humans may have had a far more ancient multiregional origin, in addition to which they estimate that periods of around two million years are required for the production of sufficient genetic distances to represent speciation (Cunroe and Thorne, 2004).

iv) Criticisms of Fuerle's arguments.

Fuerle's arguments must be dismissed on the basis of his having used incompatible measures of genetic distance based on the analysis of different genes. Measures of genetic distance involving mtDNA and SNP's for example are simply inappropriate for the purposes of inter-specific comparison as the different genes involved will have been subject to markedly different selection pressures and are not likely to have diverged at the same time. This is why alternative estimates of the diversity between the gorilla species and the common chimpanzee and bonobo, were listed by this author based on various nuclear loci and autosomal DNA. The much higher numbers reflect the extreme variation that can be expected when different genes are considered. Fuerle's presentation of the data is also misleading for another reason, namely he makes no mention of the current debates surrounding gorilla and chimpanzee/bonobo taxonomy; as new research on these taxa regularly generates novel and in some cases wildly variable estimates of genetic distance between these primates (*see eg.* Thalmann *et al.*, 2006). Finally, it would appear that Fuerle might actually have exaggerated the inter-racial genetic distances by an order of magnitude. The genetic distance between African and Australopapuan populations for example, is given as 0.03 in Cavalli-Sforza (1991) and Cavalli-Sforza *et al* (1994), however Fuerle, citing Salter (2003), lists the difference as 0.33. This may be due to differences in the way in which these estimates were calculated which in turn weakens the case for meaningful inter-specific comparisons even further as this represents yet another way in which the measures lack standardization.

v) Anecdotal evidence.

A small number of isolated observations have been made which could be interpreted as being supportive of the two biological species hypothesis. Sarich and Meile have suggested that the racial differences in craniofacial morphology are typically around ten times the corresponding differences between the sexes within a given race, which they note, is larger than the comparable differences that taxonomists use in distinguishing common chimpanzees from bonobos (Sarich and Meile, 2004). Additionally Coppinger and Schneider have suggested that the degree of mtDNA difference between dogs/wolves and coyotes (two species – *Canis lupus* and *C.latrans*) has been found to be less than that observed between the various human racial groups (Coppinger and Schneider, 1995).

As was mentioned previously, these findings are based on isolated observations primarily of anecdotal interest, and taken together provide only very minor support for the two biological species hypothesis. Whereas, within the context of conventional taxonomy, there appears to be good evidence suggestive of a subspecies-like population structure within *H.sapiens,* there appears to be little substantial evidence for differentiation at the level of biological species.

Discussion.

i) Summary of findings.

In answer to the main question of this thesis, there are grounds for suggesting the reality of currently unrecognized extant taxa within the *H.sapiens*: this argument is based upon the following lines of reasoning. Firstly, as has been demonstrated there exists a considerable degree of diversity (as measured by heterozygosity) within the taxon *H.sapiens sapiens*, suggestive of the existence of around five major clades corresponding to unacknowledged subspecies, which consistently manifest themselves despite differences in the genetic markers used. And secondly, as the phylogenetic species concept does not recognize the validity of subspecies as a division, opting instead to label the least inclusive taxon as 'species', a case could be made for the minor clades within the genus *Homo* qualifying as phylogenetic species in their own right, especially when considered in light of the evidence suggestive of the idea that lineage admixture between human racial groups is in fact fairly peripheral and is not negating the evolutionary distinctiveness of those groups. The case for the two biological species hypothesis is extremely weak. Aside from some largely anecdotal observations, arguments based on the use of comparative genetic distances between biological species that are suggestive of the idea that the distances between major racial groups within *H.sapiens* are greater than the distances recorded between certain other primate species and *H.erectus*; collapse on the basis that such comparisons have been made by illegitimately comparing F_{ST} estimates derived for different genes with different potential selection histories.
The observation that *H.sapiens* is fairly unique amongst primates in that the majority of the variation lies

within the racial groups rather than between them is suggestive of the idea that infra-specific variation measured using the classical race/subspecies concept would be expected to be low, however the diversity of phylogenetic species would be expected to be high. The data would appear to confirm this as cladistic analysis reveals the existence of a population structure comprising between four and five subspecies (major clades), however if the minor clades are used in estimating the numbers of phylogenetic species, then the number could be as high as 38.

ii) Closing remarks.

Considerable amounts of time and energy have been invested by palaeoanthropologists in an effort to identify human ancestors from the fossil record, the same is true in the case of hominologists who have expended similar resources in an effort to locate cryptid hominid and hominoid species such as the Alma and the Sasquatch, however these same researchers seem reluctant to suggest that the extant genus *Homo* may contain taxa in need of identification and classification. This is of course entirely understandable as the issue of race is often a politically incendiary one and researchers wish to maintain their careers and reputations, but is it necessarily wise to ignore the reality of human taxonomic diversity?

There exists to the mind of this author, three salient reasons why the recognition of new extant human taxa is desirable. Firstly, it would appear that only through consideration of the timing and causes of the points of divergence between the major taxa of humanity can an entirely accurate model of human evolution be devised. Secondly, medicine needs to move towards a greater appreciation of the role played by individual and group differences in disease susceptibility and response to treatment. Only through an accurate knowledge of the ways in which human biodiversity is apportioned can medicine realize this goal and save the maximum number of lives in the future.

Finally a normative case needs to be considered, as the recognition of new extant human taxa should in no way effect the fundamental human rights of every living person. In point of fact it could be argued that the recognition of human taxonomic diversity within the context of a universal human rights framework provides a good legalistic basis for extending those same human rights to potential relict hominid populations (such as Orang-pendek and the Almas) should they a) be demonstrated to exist and b) be demonstrated to be distinct species or subspecies within the genus *Homo*.

References.

Adcock, G. J., Dennis, E. S., Easteal, S., Huttley, G. A., Jermiin, L. S., Peacock, W. J., and Thorne, A. (2001). "Mitochondrial DNA sequences in ancient Australians: Implications for modern human origins." *Proceedings of the National Academy of Sciences*, 98:537-542.

Becquet, C., Patterson, N., Stone, A. C., Przeworski, M., and Reich, D. (2007). "Genetic structure of chimpanzee populations." *PLoS Genetics,* 3: e66, doi:10.1371/journal.pgen.0030066.

Bowcock. A. M., Ruiz-Linares, A., Tomfohrde, J., Minch, E., Kidd, J. R., and Cavalli-Sforza, L. L. (1994). "High resolution of human evolutionary trees with polymorphic microsatellites." *Nature*, 368:455-457.

Cann, R. L., Stoneking, M., and Wilson, A. C. (1987). "Mitochondrial DNA and human evolution." *Nature,* 325:31-36.

Caramelli, D., Lalueza-Fox, C., Vernesi, C., Lari, M., Casoli, A., Mallegni, F., Chiarelli, B., Dupanloup, I., Bertranpetit, J., Barbujani, G., and Bertorelle, G. (2003). "Evidence for a genetic discontinuity between Neandertals and 24,000-year-old anatomically modern Europeans." *Proceedings of the National Academy of Sciences*, 100:6593-6597.

Cartwright, S. A. (1857). *Natural History of the Prognathous Species of Man*. New York Day, New York.

Cavalli-Sforza, L. L. (1991). *Genes, Peoples and Languages*. North Point Press, San Francisco.

Cavalli-Sforza, L. L., Menozzi, P., and Piazza, A. (1994). *The History and Geography of Human Genes*, Princeton University Press, New Jersey.

Coon, C. S. (1962). *The Origins of Races*. Alfred A. Knopf, New York.

Coppinger, R., and Schneider, R. (1995). "Evolution of working dogs." In: Serpell, J. (ed). *The Domestic Dog: Its Evolution, Behaviour and Interactions with People*. Cambridge University Press, Cambridge. pp. 21-47.

Culver, M., Johnson, W. E., Pecon-Slattery, J., and O'Brien, S. J. (2000). "Genomic ancestry of the American puma (*Puma concolor*)." *Journal of Heredity*, 91:186-197.

Curnoe, D., and Thorne, A. (2003). "Number of ancestral human species: a molecular perspective." *HOMO*, 53:201-224.

Dart, R. A. (1953). "The predatory transition from ape to man." *International Anthropological and Linguistic Review*, 1:201-217.

Delgado, R., and Stefancic, J. (2001). *Critical Race Theory: An Introduction*. New York University Press, New York.

Dobzhansky, T. (1970). *Genetics of the Evolutionary Process*. Columbia University Press, New York.

Edwards, A. W. F. (2003). "Human genetic diversity: Lewontin's fallacy." *BioEssays*, 25:798-801.

Eizirik, E., Kim, J., Menotti-Raymond, M., Crawshaw, P. G. Jr., O'Brien, S. J., and Johnson, W. E. (2001). "Phylogeography, population history and conservation genetics of jaguars (*Panthera onca*, Mammalia, Felidae)." *Molecular Ecology*, 10:65-79.

Eldridge, N., and Cracraft, J. (1980). *Phylogenetic Analysis and the Evolutionary Process: Method and Theory in Comparative Biology*. Columbia University Press, New York.

Forbes, S. H., Hogg, J. T., Buchanan, F., Crawford, A., and Allendorf, F. (1995). "Microsatellite evolution in congeneric mammals: domestic and bighorn sheep." *Molecular Biology and Evolution*, 12:1106-1113.

Fuerle, R. D. (2008). *Erectus Walks Amongst Us*. Spooner Press, New York.

Garcia-Moreno, J., Matocq, M., Roy, M., Geffen, E., and Wayne, R. K. (1996). "Relationships and genetic purity of the endangered Mexican wolf based on analysis of microsatellite loci." *Conservation Biology*, 10:376-389.

Gee, H. (2004). "Flores, God and cryptozoology." *Nature,* 431:1055-1061.

Girman, D. J., Vil, C., Geffen, E., Creel, S., Mills, M. G. L., McNutt, J.W., Ginsberg, J., Kat, P. W., Mamiya. K. H., and Wayne. R. K. (2001). "Patterns of population subdivision, gene flow and genetic variability in the African wild dog (*Lycaon pictus*)." *Molecular Ecology,* 10:1703-1723.

Goodrum, J. (2002). "The race FAQ." Available at: http://www.goodrumj.com/RFaqHTML.html (Retrieved on 24/11/08).

Guillén, A., Barrett, G. M., and Takenaka, O. (2005). "Genetic diversity among African great apes based on mitochondrial DNA sequences." *Biodiversity and Conservation,* 14: 2221-2233.

Gutiérrez, G., Sánchez, D., and Marín, A. (2002). "A reanalysis of the ancient mitochondrial DNA sequences recovered from Neandertal bones." *Molecular Biology and Evolution,* 19: 1359-1366.

Heuvelmans, B. (1969). "Note préliminaire sur un spécimen conservé dans la glace, d'une forme encore inconnue d'Hominidé vivant: *Homo pongoides* (sp. seu subsp. nov.)." *Bulletin de l'Institut Royal des Sciences Naturelles de Belgique,* 45:1–24.

Heuvelmans, B., and Porshnev, B. F. (1974), *L'Homme de Néanderthal est Toujours Vivant.* Plon, Paris.

Hooton, E. A. (1926). "Methods of racial analysis." *Science,* 63:75–81.

Jensen-Seaman, M. (2000). "Western and eastern gorillas: estimates of the genetic distance." *Gorilla Journal,* 20:21-23.

Jones, T., Ehardt, C. L., Butynski, T. M., Davenport, T. R. B., Mpunga, N. E., Machaga, S. J., and de Luca, D. W. (2005), "The highland mangabey *Lophocebus kipunji*: a new species of African monkey." *Science,* 308:1161-1164.

Jorde, L., Rogers, A., Bamshad, M., Watkins, W.S., Krakowiak, P., Sung, S., Kere, J., and Harpending, H. (1997). "Microsatellite diversity and the demographic history of modern humans." *Proceedings of the National Academy of Sciences,* 94:3100-3103.

Krantz, G. S. (1986). "A species named from footprints." *Northwest Anthropological Research Notes,* 19:93-99.

Krantz, G. S. (1992). *Big Footprints: A Scientific Inquiry Into the Reality of Sasquatch.* Johnson Books, Boulder.

Krings, M., Stone, A., Schmitz, R. W., Krainitzki, H., Stoneking, M., and Pääbo, S. (1997). "Neandertal DNA sequences and the origin of modern humans." *Cell,* 90:19-30.

Kyle, C. J., and Strobeck, C. (2001). "Genetic structure of North American wolverine (*Gulo gulo*) populations." *Molecular Ecology,* 10:337-347.

Leiberman, L., Hampton, R. E., Littlefield, A., and Hallead, G. (1992). "Race in biology and anthropology: a study of college texts and professors." *Journal of Research in Science Teaching,* 29:301-321.

Lewontin, R. C. (1973). "The apportionment of human diversity." *Evolutionary Biology,* 6:381-397.

Long, J. C., and Kittles, R. A. (2003). "Human genetic diversity and the nonexistence of biological races." *Human Biology*, 75:449-471.

Mayr, E. (1969). *Principles of Systematic Zoology*. McGraw-Hill, New York.

Meldrum, D. J. (2004). "Midfoot flexibility, fossil footprints, and Sasquatch steps: new perspectives on the evolution of bipedalism." *Journal of Scientific Exploration*, 18:67-79.

Mishler, B. D., and Theriot, E. C. (2000). "The phylogenetic species concept (*sensu* Mishler and Theriot): monophyly, apomorphy, and phylogenetic species concepts." In: **Wheeler, Q., and Meier, R.** (Eds). *Species Concepts and Phylogenetic Theory: A Debate*. Columbia University Press, New York. (Three chapters) pp. **44-54, 119-132, 179-184.**

Morton, N. E., Chung, C. S., and Mi, M. P. (1967). *Genetics of Interracial Crosses in Hawaii*. Karger Monographs in Human Genetics, Vol. 3, Basel.

Nei, M., and Roychoudhury, A. (1993). "Evolutionary relationships of human populations on a global scale." *Molecular Biology and Evolution*, 10:927-943.

Paetkau, D., Waits, L., Clarkson, P., Craighead, L., Vyse, E., Ward, R., and Strobeck, C. (1998). "Variation in genetic diversity across the range of North American brown bears." *Conservation Biology*, 12:418-429.

Paetkau, D., Amstrup, S. C., Born, E. W., Calvert, W., Derocher, A. E., Garner, G. W., Messier, F., Stirling, I., Taylor, M. K., Wiig, Ø., and Strobeck, C. (1999). "Genetic structure of the world's polar bear populations." *Molecular Ecology*, 8:1571-1584.

Polziehn, R. O., Hamr, J., Mallory, F. F., and Strobeck, C. (2000). "Microsatellite analysis of North American wapiti (*Cervus elaphus*) populations." *Molecular Ecology*, 9:1561-1576.

Porshnev, B. F. (1963). *The Present State of the Problem of Relic Hominoids*. (in Russian). Viniti, Moscow.

Porshnev, B. F. (1969). "The problem of relic *Paleanthropus*." (in Russina). *Sovetskaya Etnografia*, 2:115-30.

Porshnev, B. F. (1974). "The Troglodytidae and the Hominidae in the taxonomy and evolution of higher primates." *Current Anthropology*, 15:449-450.

Reinartz, G. E., Karron, J. D., Phillips, R. B., and Weber, J. L. (2000). "Patterns of microsatellite polymorphism in the range-restricted bonobo (*Pan paniscus*): considerations for interspecific comparison with chimpanzees (*P. troglodytes*)." *Molecular Ecology*, 9:315-328.

Rushton, J. P. (2005). "Ethnic nationalism, evolutionary psychology and genetic similarity theory." *Nations and Nationalism*, 11:489-507.

Salter, F. K. (2003). *On Genetic Interests: Family, Ethny, and Humanity in an Age of Mass Migration*. Peter Lang, Frankfurt.

Sarich, V., and Miele, F. (2004). *Race: The Reality of Human Differences*. Westview Press, New York.

Schwartz, M.K., Mills, L. S., McKelvey, K. S., Ruggiero, L. F., and Allendorf, F. W. (2002). "DNA reveals high dispersal synchronizing the population dynamics of Canada lynx." *Nature*, 415:520-522.

Simpson, G. G. (1984). "Mammals and cryptozoology." *Proceedings of the American Philosophical Society*, 128:1-19.

Stringer, C., and Andrews, P. (1988). "Genetic and fossil evidence for the origin of modern humans." *Science*, 239:1263-1268.

Templeton, A. R. (1998). "Human races: a genetic and evolutionary perspective." *American Anthropologist*, 100:632–650.

Thalmann, O., Fischer, A., Lankester, F., Pääbo, S., and Vigilant, L. (2006). "The complex evolutionary history of gorillas: insights from genomic data." *Molecular Biology and Evolution*, 24, 146-158.

Thorne, A., and Wolpoff, M. (2003). "The multiregional evolution of humans." In: Smith, F., and Spencer, F (eds). *Scientific American Special Editions – The Origin of Modern Humans*. Scientific American, New York. pp. 76-93.

Trinkaus, E., and Shipman, P. (1983). *The Neandertals: Changing the Image of Mankind.* Alfred A. Knopf Inc, New York.

Uphyrkina, O., Johnson, W. E., Quigley, H., Miquelle, D., Marker, L., Bush, M., and O'Brien, S. J. (2001). "Phylogenetics, genome diversity and origin of modern leopard, *Panthera pardus*." *Molecular Ecology*, 10:2617-2633.

Van Hooft, W. F., Groen, A. F., and Prins, H. H. T. (2000). "Microsatellite analysis of genetic diversity in African buffalo (*Syncerus caffer*) populations throughout Africa." *Molecular Ecology*, 9:2017-2025.

Waits, L., Taberlet, P., Swenson, J. E., Sandegren, F., and Franzén, R. (2000). "Nuclear DNA microsatellite analysis of genetic diversity and gene flow in the Scandinavian brown bear (*Ursus arctos*)." *Molecular Ecology*, 9:421-431.

Walker, C. W., Vil, C., Landa, A., Lindén, M., and Ellegren, H. (2001). "Genetic variation and population structure in Scandinavian wolverine (*Gulo gulo*) populations." *Molecular Ecology*, 10:53-63.

Wallace, R. B., Gómez, H., Felton, A., and Felton, A. M. (2006). "On a new species of titi monkey, genus *Callicebus* Thomas (primates, Pitheciidae), from Western Bolivia with preliminary notes on distribution and abundance." *Primate Conservation*, 20:29-39.

Wilton, A. N., Steward, D. J., and Zafiris, K. (1999). "Microsatellite variation in the Australian dingo." *Journal of Heredity*, 90:108-111.

Wise, C., Sraml, M., Rubinsztein, D., and Easteal. S. (1997). "Comparative nuclear and mitochondrial genome diversity in humans and chimpanzees." *Molecular Biology and Evolution*, 14:707-716.

THE MYSTERY CRYPTIC ANIMAL OF NORTHERN NEW SOUTH WALES
Gary Opit

Northern New South Wales has its own mystery animal stalking the streets and bushland. Many local people have seen it. When talking to friends they are often surprised to find that they too have seen it. It looks like someone has crossed a dog with a kangaroo. It is observed at night as locals drive their cars through the wildlife corridors that surround our homes. Fishing parties on have seen it, as have people relaxing on their verandah. Families pedaling their bicycles during the day have seen it. At first, they believe that the animal is a dog or fox. Closer observations reveal unexpected characteristics. "What on Earth is this?" they ask themselves.

Families have their own names for it. The *Devil Dog,* the *Hound from Hell,* the *Ocean Shores Oddity,* the *Billinudgel Beast,* the *Mullumbimby Monster* and the *Byron Beast.* Every Wednesday morning at 6.20 am on 94.5 FM North Coast Regional ABC Radio you can listen and speak with me on the Wildlife Talk-back broadcast. I talk about the seasonal behaviour of local wildlife and identify fauna species for listeners from their descriptions of physical features or calls. Over the last 10 years I have received many reports of these animals and some have made the local news service on radio, TV and newspapers.

Perhaps these animals are just mangy foxes or wild dogs. Farmers and other rural dwellers regularly observe these introduced species. However, the descriptions sound more like an animal that is supposed to be extinct. The strange, waddling gait, the kangaroo-like tail and the brown bands across the back remind us of the remarkable thylacine. Like the koala and the kangaroo, the thylacine is a unique Australian marsupial with the female rearing two or three pups in a backward-opening pouch. It differs in that it hunts other animals for its food.

Believed to be extinct in Australia for perhaps 3000 years, it continued to survive in Tasmania until 1936 when the last captive thylacine died. Known there as the Tasmanian Tiger or wolf, it succumbed to hunting, habitat destruction and perhaps introduced diseases. It was feared that it may attack livestock but a recent study of the detailed records kept by the big sheep stations in Tasmania, listing the cause of all sheep deaths, found almost no evidence that the thylacine ever attacked domestic animals. It fed almost exclusively on small bushland animals such as wallabies, bandicoots and bush rats.

Since its supposed extinction there have been hundreds of reported sightings in both Tasmania and the Australian mainland. Some controversial photographs have been taken but no definite evidence has been forthcoming to prove the animal still exists. Scientists at the Australian Museum have been trying to clone a thylacine from a juvenile preserved in alcohol.

The reports that I have collected appear to describe the survival of a small number of these animals in the rugged wilderness of the Whian Whian, Nightcap and the Border Ranges. The theory is that over the years, the population has increased and now they are being observed in the coastal nature reserves. Like the Whian Whian Oak, the Wollemi Pine and other supposedly extinct species, there is a possibility that a most wonderful Australian has returned. Perhaps you will be the first person to photograph this animal and prove that it exists. If it is the thylacine then it should not be harmed as it an endangered species. It is even possible that you could find a dead thylacine, as there has been the occasional report of such an animal lying on the side of the road, the victim of a vehicle impact. Such specimens, if found, should be taken to the national parks service for identification.

It is not a dangerous animal and early last century in Tasmania it was kept just like a pet dog. Ancient cave paintings in Kakadu illustrate thylacines carrying dilly bags around the neck so it was a companion of Aboriginal people before the dingo arrived from South-east Asia. Because it is a carnivore, it is naturally cryptic, hiding in the vegetation to spring out onto small animals. It lives in small family groups that range widely over large territories. Perhaps it is long extinct and people are only seeing mangy foxes or dogs. Keep your eyes open & if you think that you have observed something unusual write down a description noting the date, time and place or you can contact me at garyopit@bigpond.com.

The 50 Reports

I received the following reports during my Wildlife Talkback radio broadcasts over the last ten years or from people contacting me directly. The witness was going about their normal lives, driving back & forth to work, taking children to school, or off to the shops. Most only observed the animal on a single occasion even though they may have travelled the road countless times. The sighting of a single individual is usual though two reports include a pair of animals and a female with young following her. Most of the observations were from cars though some witnesses were walking or bicycle riding.

Most of these sightings only lasted for a minute or two, as the animal crossed a road or paddock. Most observations have been at night though others have been during daylight & all were close enough, usually only a few metres away, to enable a very good description. As is normal practise, most were not carrying cameras with them in the hope that something remarkable would occur worthy of photography. If

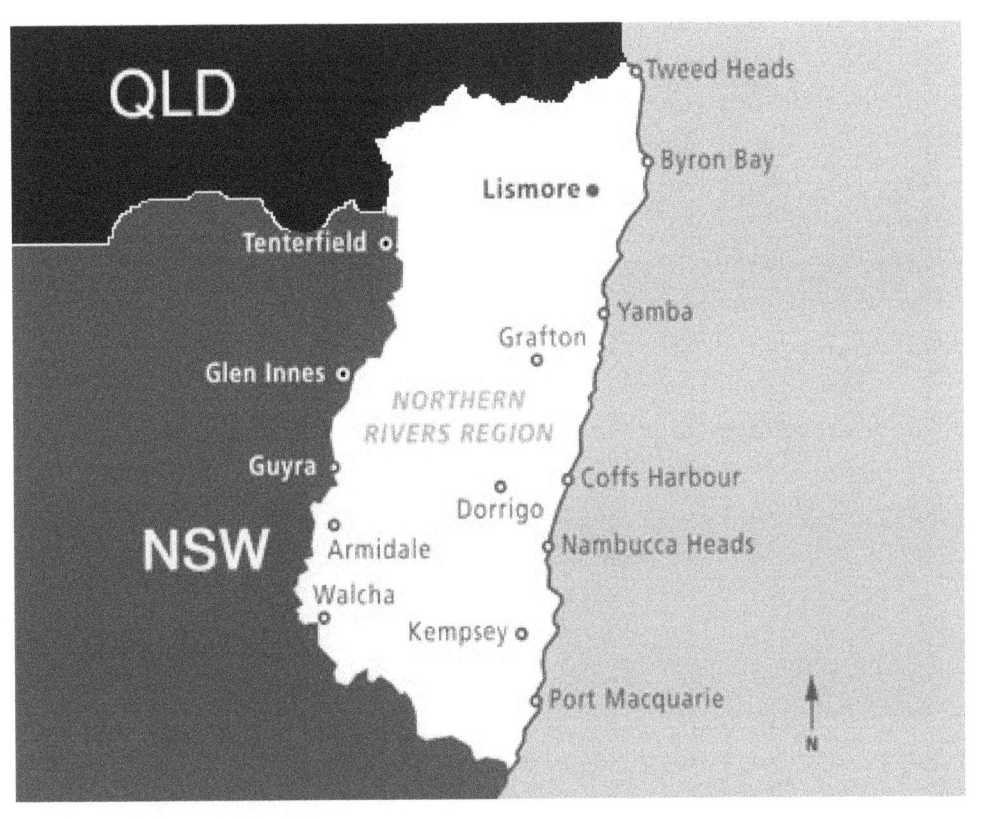

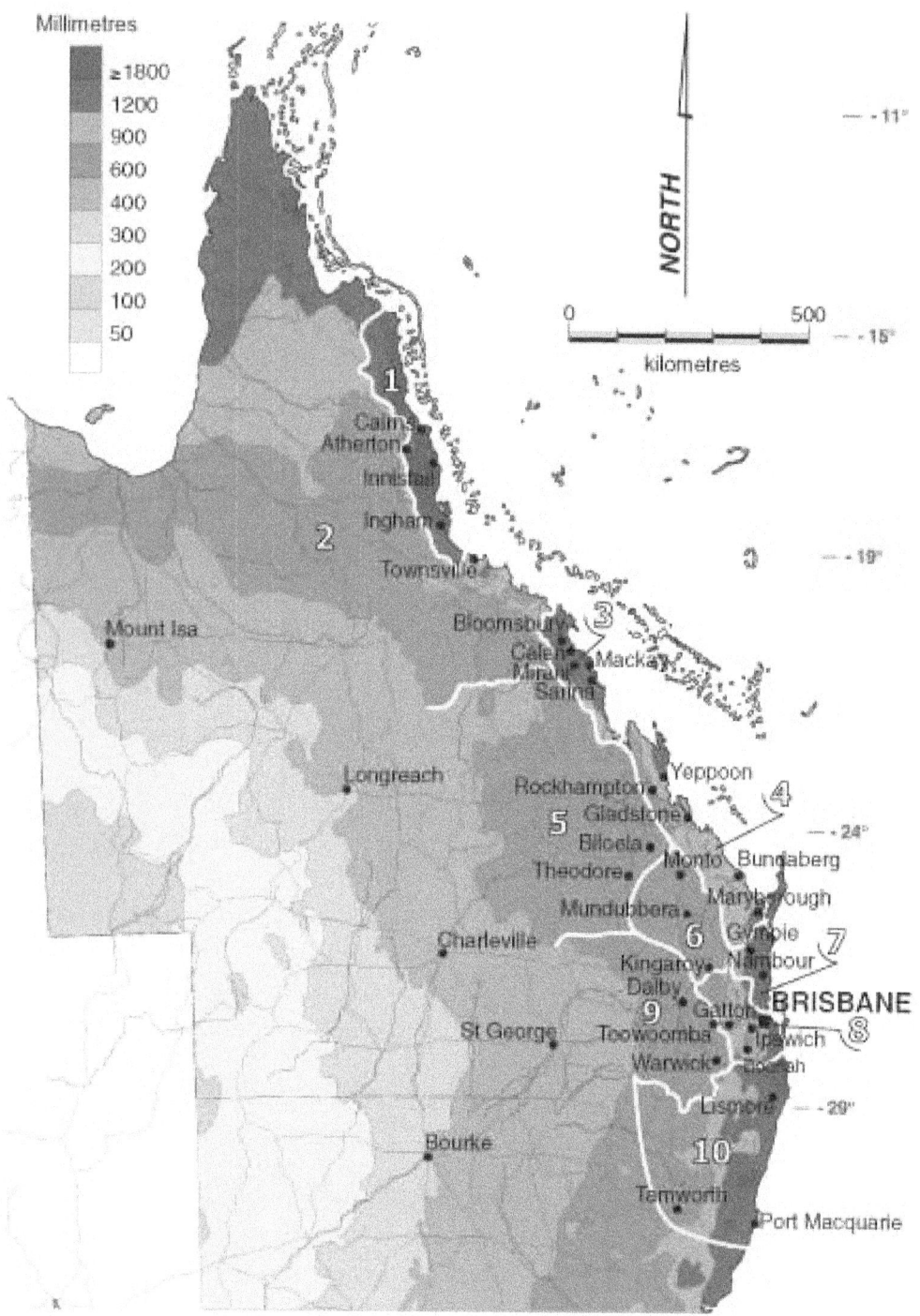

they did happen to have a camera, the witness would at first think it was not an interesting subject to photograph until they realised how unusual it was and then found that they did not have enough time to retrieve the camera before the animal moved off. However, many witnesses have stated that they now carry cameras with them just in case they view the animal once more.

Many did not know of the existence of the thylacine & believed that it was some freak of nature, wherein someone had hybridized a dog with a kangaroo. Others recognised it as an animal that they had previously seen as a photo or drawing in a newspaper, magazine, books or on TV, but generally could not remember the name of the animal or how rare or unusual it was. However, some witnesses were very well acquainted with native plants & animals & were amazed to observe an animal that they were positive did not exist in this locality.

Some of these reported sightings may indeed be of a thylacine. However, others may well be sightings of wild dogs or mangy foxes. At least one witness has described seeing very pale coloured stripes across the back of the animal at close range, that were not, at first notice, visible. This may explain why many sightings of animals that look very much like a thylacine do not appear to have stripes across the back.

May 1964 Monday 5 am in Whian Whian state forest.

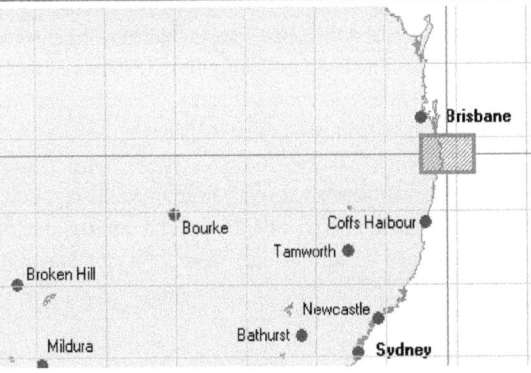

Clive gave me a very detailed description of the animal observed twice in 1 week. He worked for 5 years with Standard Saw Mills of Lismore as a logger and often saw dingos. Clive said that there were no foxes up there in the Whian Whian Range. He only saw the Thylacine twice during the same week, approximately 3 years after he began working in those ranges. He was camped at the old army hut and was driving to the logging coup on the western side of Peach Mountain lookout at 5 am on a Monday morning when he observed a thylacine cross the road 3 metres in front of his car. It had distinctive stripes across its back & rump, which sloped down to a long kangaroo-like tail. It came from the left hand side up the slope, crossed the road & then leapt up the bank. He observed it again at 5 pm on the Friday afternoon of the same week whilst driving back from work at exactly the same place. This time it was coming down the slope from the right, jumped down the bank onto the road & continued down the slope as if it regularly used the same animal track. He never saw the animal again even though he drove those same roads for another 2 years.

1970, Crabbs Creek. Schoolteacher Mark was working on a banana farm during the school holidays and as they descended from a forested ridge top at the end of the day, the owner's German Shepard dog began growling at something sheltering within an old, partly collapsed banana-packing shed overgrown with vines. The dog rushed in to attack the animal & Mark, the farm owner and several other workers were surprised to see the dog backing out of the shed with an animal almost as large covered with brown stripes across its back and a thick, stiff, kangaroo-like tail. The strange animal had a huge jaw that opened to an extent, greater than the dog, and it gave forth with a bizarre cry unlike anything that they had heard before.

The farm owner yelled out "It's a monster, we will have to kill it" and picking up a stone, threw it at the strange animal. The stone missed its mark and the animal, looking up, saw the people and ran at great speed up the slope with a very unusual gait. The dog and the people chased the animal into a large hollow log where it crouched to stare at them. The owner remarked that they would have to kill the animal as he would not allow a monster to live on the farm. Then they all descended back through the bananas to head for home. The next day the farm owner brought up his rifle but the animal was gone and they never saw it again.

Tyagarah 1979, David saw a strange animal, which reminded him of the "Tyagarah Lion" which he had heard about over the years.

1982 at Lake Ainsworth, Lennox Head, Grey's Lane, Tyagarah, Uki and Terrania Creek "Rabbit" observed 5 times a thylacine-like animal with a striped rump, always around 4 am when driving before first light on his delivery rounds.

1988, Cawongla near Kyogle on the roadside at night. Len saw a thylacine-like animal showing distinct dark brown banding on the rump, hips, legs & along the tail. The tail was thickly furred which reminded him of a photo of a numbat. The bands were about 2 cm wide & about 6 cm apart. The front paw was lifted up near the snout. The snout & the tail were held straight & the ears were cocked up.

1989, Terania Creek Road, The Channon, running across the road at night in front of his car, Peter saw a thylacine-like animal showing distinct dark brown banding across the body. The tail was thickly furred. Following this animal were 3 smaller identical animals. This is the only report of a mother and its young.

1992 Ewingsdale, Tony saw a creature on a bright and sunny mid-morning 50 metres away that he was at a total loss to identify. It had a long thin straight tail, short sandy brown fur, a greyhound look to it and an odd gait. The animal was not concerned and it headed towards a large fig tree where friends let their chooks out most days. It disappeared behind the fig tree and did not re emerge. It never stopped and kept a constant pace. The area was open paddock with a ridge-line that the animal was moving along.

1992 Coopers Shoot Road, Bangalow, 8.30 pm Vicki said that the weirdest animal appeared in the middle of the road. She had to slam the brakes on to prevent hitting it. The animal then snarled at us showing long pointed teeth, before disappearing into long grass. The color of this animal was light in appearance & there were no stripes. She and a work mate looked at each other in total disbelief & both said together "What in god's name was that? She stated "We both knew that this was neither dog nor fox." When explaining the incident later to family & friends she could best describe it as looking something like a Tasmanian tiger. Every one laughed at her and cried in disbelief "A Tasmanian tiger in Byron Bay!"

1995, Coopers Lane, Main Arm, Hayley observed a golden-fawn individual with a striped tail.

16th November, 1997, Sunday, 7-30 a.m. at Lennox Head, between Seven Mile Beach and Lake Ainsworth, near Camp Drew. Paul and his partner observed from their car, only 1 metre away, a dog-sized animal with black stripes down its back and rump with one stripe across the base of the tail which was long and stuck straight out behind it. It was covered in short sandy fur, with a long thin head and face with upright ears and he was certain that it was not a cat, fox or dog. He phoned the national parks service, a local wildlife carer and some time later, the north coast ABC radio station while I was broadcasting my Wildlife Talkback programme.

18 November 1997, 9 a.m. North Tumbulgum, adjacent Hogan's Rainforest Nature Reserve on the NSW / Qld border. Jan and two other family members observed on their property a striped dog-like animal with a head almost like a kangaroo and stripes continuing onto the long stiff tail. They had previously observed it on two earlier separate occasions and enquiring of the neighbours, were told that all three families on adjoining properties had observed the animal going back at least ten years but had never bothered to report it.

November 1997, Upper Durobby Creek, in the foothills of the MacPherson Ranges. Dennis, a neighbour of Jan, phoned to describe a similar animal. Of particular interest was that the animal that he saw had no stripes on the body, though it did have pale bands along the tail. It was greyhound-like, the head was like that of a kangaroo, particularly because of its kangaroo-like ears that stood straight up, the ribs were tucked up and the rump was uplifted.

The tail was thick at the base and as long as the body, was round in cross section and went to a point. It was shaped like a kangaroo's tail, but held straight out behind instead of dragging on the ground. Its fur was very short, about 15 mm long, of a greyish to light brown colour and was not at all mangy. At the base of the tail there was an orange ring about 50 mm wide and it was followed by 6 to 8 yellow rings about 40 mm wide.

His wife was the first of the family to observe the animal, 3 months before while taking their child to the bus stop down their one kilometre long driveway. Two days later their teenage son observed it and described it as being a cross between a kangaroo and a greyhound. Two weeks later Dennis finally saw it sitting on the roadside while driving down their road to pick up their child from the bus stop at 4 pm. He watched it walk off from beside his car for one hundred metres. It walked like a dog and its unusual tail did not look out of proportion with its body. He stated that it didn't look like some unusual hybrid but a species of carnivorous marsupial. His teenage son has since set up a large live animal trap in an attempt to capture it.

1997, Mt Warning, Heidi described her brother's observation of a thylacine.

He and his friend were quite close to the base of the mountain when they passed an odd looking animal. He said when they stopped it ran out onto the road behind them and froze for a moment also looking at them before dashing off into the forest. He said it looked exactly like the Tasmanian tigers he had seen pictures of in books except that it was a lot darker in colouring. He said it had stripes & a sloping back.

1997, 4am, Uki. Peter described his sighting of 1997 on the Murwillumbah Road to Uki near the intersection to Mount Warning at 4am when he observed what at first he thought was a fox on the side of the road. He immediately noticed that it had an unusual shuffling walk with the rear of the body sloping backwards & thought that it had a dislocated hip, which he had observed on an injured dog previously. He expected to catch up with the animal with ease because of its disability but was surprised that when it became aware of his cars' approach it raced off along the roadside with incredible speed. He accelerated up to it and observed that its back, rump & tail were covered with dark bands. He was sure that it definitely was not a fox or a dog, that its snout was not pointed like a fox and that it had distinctive rounded ears. It ran off into the vegetation adjacent the road.

1999, Brunswick Heads Jodi saw a striped animal cross the road between the fish co-op and the highway.

1999, Federal Graham and Rosalind had a close view of a strange animal when driving between Whian Road and Bates Road near the old Dip when the headlights illuminated it. The size of a dog, it had a big distinctive head, brown fur and a stiff kangaroo-like tail. It had very obvious stripes across the back and the base of the tail which blended in with the brown fur so that the stripes would not be so visible from further away. It had a very unusual manner of walking quite unlike a dog. They decided that it could only be a Tasmanian tiger and phoned the national parks service to report the sighting and where most annoyed when they were not believed and told that they could not possibly have seen such an animal because it is extinct. Their daughter also saw it a year later and described the same animal. A friend, Eric

Cornwall told them that he had the same kind of animal cross the road in front of his car near Wooyung and although he applied the breaks he hit the animal. He stopped and had a look but the animal had run off unhurt.

2000, Mahers Lane, Terranora Don saw a long thin dog-like animal with stripes across its back & the base of its long thin tail in the evening as he drove down the road from his home through farmland. It was unconcerned by his approaching car and he was able to get a close look at it. Some years later he saw it again at the same time and place & then his wife saw it in similar circumstances near their house. Nearby residents Allan & Maureen also observed the animal.

15 January 2003, 9.30 am, Stock Route Road, Billinudgel, Mailman Peter drove right up to a strange looking animal standing on an earth bank on the southern section of Stock Route Road, just behind Billinudgel. As tall as a medium-sized dog, it looked something like a whippet crossed with a kangaroo. It was covered with a fine short brown fur except for the rump and tail, which was bare skinned with individual hairs scattered evenly across it. It was completely unconcerned by the presence of his car and he closely examined it for 5 minutes before it walked off. Peter had been involved in greyhound racing for many years and so was positive that the animal was neither a dog or a fox and appeared to be a carnivorous marsupial.

2003, Wilfred Street, Billinudgel, Sue, owner of the Billinudgel Post Office, when she owned the general store next door, looked into the main street of Billinudgel one morning at 5.30 am and was surprised to see a very unusual animal standing in the middle of the road. It looked like a cross between a dog and a kangaroo. Then she found that the woman that worked in the store had recently observed two of the animals chasing and killing a swamp wallaby near her home just a few kms up the valley. A short time later a bakery representative from the Gold Coast also saw the animal & commented to Sue about it.

2003, Stock Route Road, Billinudgel, Peter, principal of Main Arm School saw what looked like a thylacine.

2003, Stock Route Road, Billinudgel Tony & Susette saw what looked like a thylacine.

2003, Stock Route Road, Billinudgel Joan Nolan saw what looked like a thylacine.

June 2003, Rosebank. Neil saw an animal that he could not identify while driving to Rosebank from Clunes at 8.30 pm. Two km to the south-west of the Green Frog Café & general store he and a friend saw in the car headlights an unusual animal cross the road 6 to 12 metres in front of them. It had a feline-like face and a long body & tail, from snout to tail tip at least one & a half metres in length, covered with yellow tawny fur.

2003, Upper Main Arm, bush regenerator Mark had a close observation of a thylacine-like animal at midday and observed its striped back and stiff tail as it stood near the roadside. Being an expert on wildlife identification he was positive that it was a thylacine. It gave a strange coughing bark-like call and bounded away.

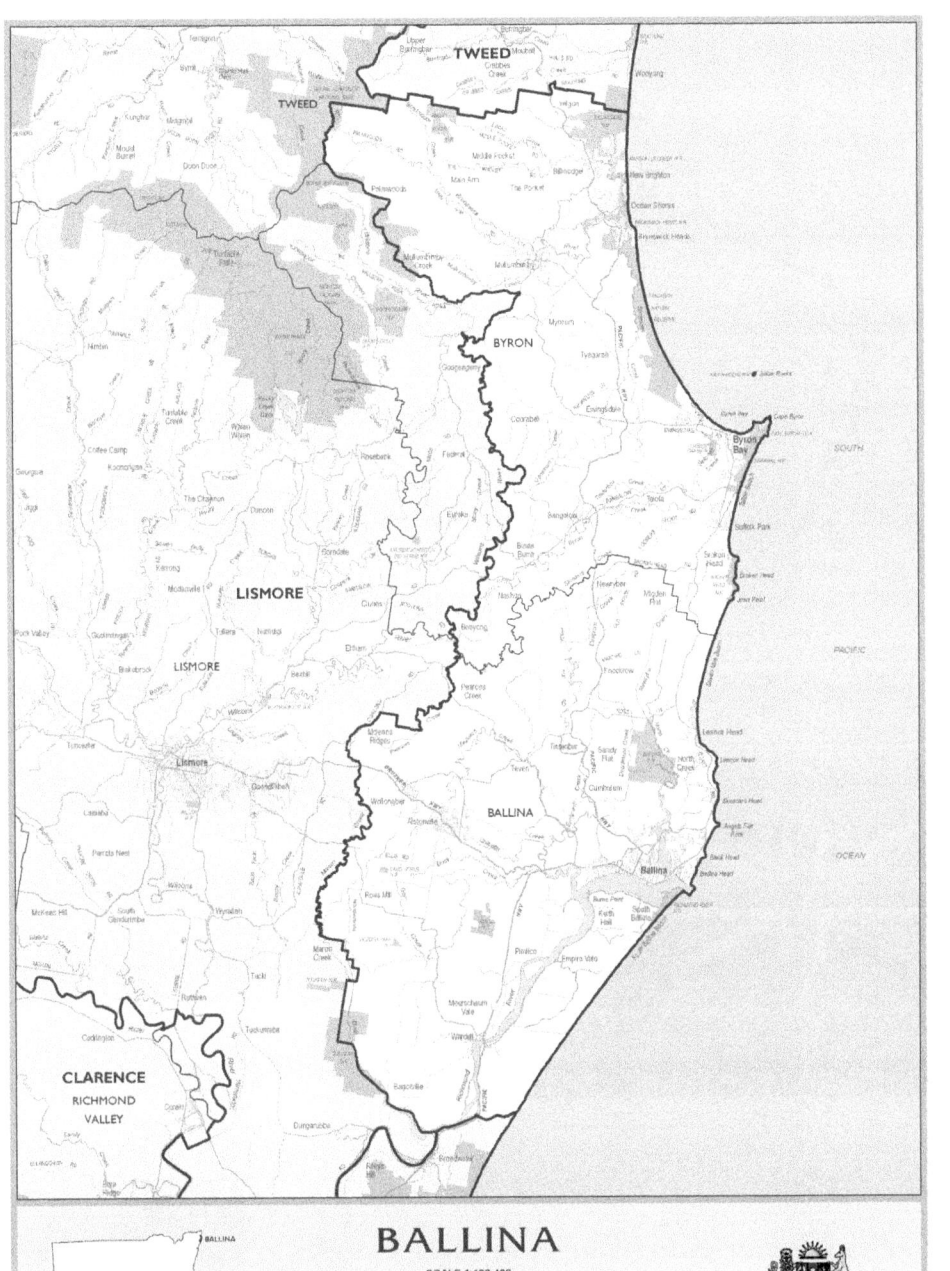

BALLINA

SCALE 1:198 400

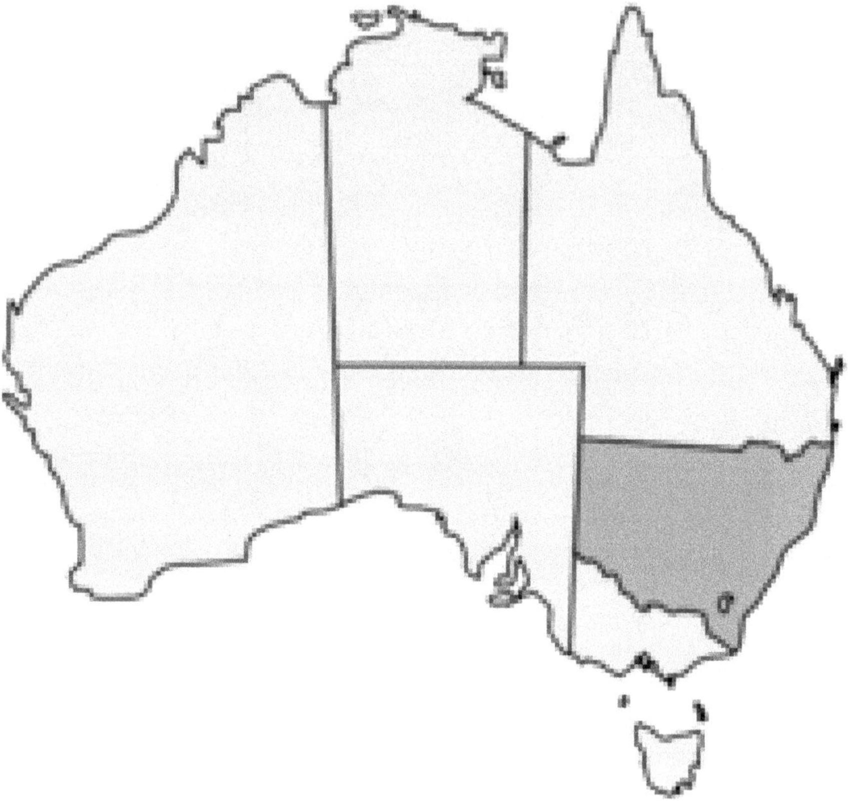

· ORTA · RECENS · QUAM · PURA · NITES

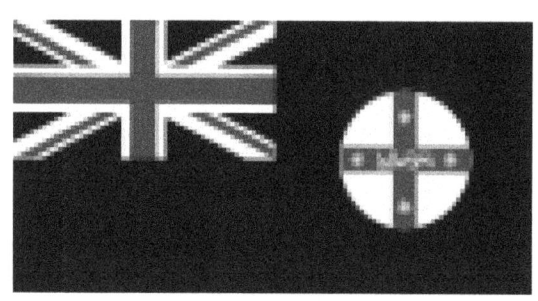

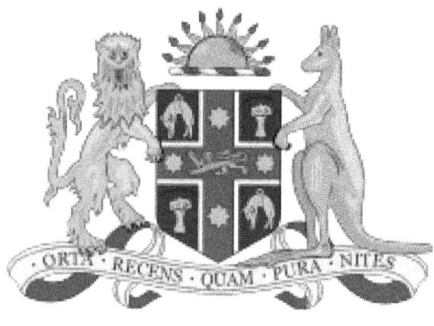

2004 Clothiers Creek Road, Cabarita, Joslyn, of Kingscliff, saw a thylacine on Clothiers Creek Road at 9.30 am as she was coming into Cabarita. It was the size of a small dingo but with an elongated, slim body and the hind quarters were more prominent. It was covered with tawny short hair with dark stripes across the back and its gait was noticeably unusual.

September 2005, 7am in the Billinudgel Nature Reserve on the trail that runs parallel to the beach several hundred metres north of the central trail entrance into the reserve. Russel noticed 2 or 3 very sleek animals slip through the bush.

8th October 2005, at approx 12pm on Shara Boulevard (not too far from the Highway). Russel saw what looked like the same animal and believe it may have been sitting as he only saw its head and shoulders in the grass but was struck by it having rounded ears.

October 2005, Terragon, south of Uki, heading towards Kunghur. Shirley woke very early to the sounds of her chooks going right off. She jumped out of bed and made her way down to the pen, where she saw an animal that was striped like a tiger that bounded over the long grass. It was thin and brownish and had a long thin tail.

November 2005, Coorabell , 9 pm Samantha saw a lion coloured creature with kangaroo like back legs hop into the bush on the Coorabell - Federal Road at the farm of the Woolnough family. She stated "I'm an artist and I paint native animals and the gait of this animal meant that it was a marsupial, not a dog or a cat. The colour was golden-fawn, 60 cm in height and the ears were rounded. It was the movement of it's pelvis and it's shyness and way it dropped it's head and pushed it's pelvis up and hopped into the bush that alerted me to the fact that this was a 'different' animal to any I'd seen before. It looked straight at me before moving away. It didn't run. I paint thylacines frequently and last year we had a gallery in Fletcher St Byron Bay with a thylacine on a rock with the word 'Imagine' inscribed on the rock. So many people, local and others, came in to describe their sightings. One of the most significant was from Maureen from Byron Post Office whose husband had spotted one while posting letters in The Pocket. She even came in with the drawing he'd done after seeing the creature. Another from a woman called Gail who worked at Durrumbul preschool and who had seen one crawl from beneath the preschool building and walk in a hopping way. She saw it for quite a while.

2005 Traegegle, Rhys observed a golden-fawn thylacine-like animal.

2005, New Brighton, Gary and Sharmaine observed a thylacine-like animal with a striped rump.

2005, South Golden Beach, Kolora Way Lyndel observed a thylacine-like animal without stripes.

7th January 2006, Saturday, 8.30 am, Drake. Greg worked at the Tenterfield Bowling Club on Friday night and he headed home at 7.45am Saturday morning. Whilst travelling east along the Bruxner Highway through the lighter wooded area, coming off the range, about ten minutes west of Drake he had a close view of a strange animal that looked like a thylacine without stripes. He reported "I can say that what I saw did not look like anything I have witnessed before. I have also heard a report about six months ago from an elderly family friend. The elderly chap's sighting was about ten years ago in the same area, closer to dawn, about 6am."

16 January 2006, Monday, 3.30 am, Anderson's Hill, Mullumbimby Mick and Fabiola where returning home along Gulgun Road adjacent Everrit's Creek crossing, 400m north of the Mullumbimby intersection and the Uncle Tom's Pies service station, when they observed a strange animal coming towards them along the eastern side of the road. It reminded them of a Fossa or a civet and definitely was not a fox, dog or cat.

Fabi was driving and so Mick was able to examine the animal closely and observed that it was 60 to 70 cm high and 1.3m long, the length of the body quite long when compared to its height. It had a very long thin tail that drooped down then lifted up towards the end. It had a large head with golden eyes and widely separated rounded ears. It was covered with short golden-fawn fur with black shadowy marks on the fur tips across the rump. Mick noticed that it had a distinct waddle of the back legs at it walked and from only 2 metres away he watched it turn away from him and saw that it had a white band at the end of the tail with a black tip. It then ran off under a barb wire fence to disappear into the regrowth vegetation.

26th January 2006, Shara Boulevard, North Ocean Shores - at around 6am (Australia Day) Russel was driving from Ocean Shores to Brunswick Heads. There was very little traffic on the roads. Towards the end of Shara Boulevard he noticed an animal walking head-on toward him, along the side of the road. He noticed the ears were rounded and that it stood about 1/2m high, its mouth was open. As he slowed down he observed it closely at an angle and was extremely surprised by the length of the tail which curved down and back up from the ground.

There were no visible markings on the body, although he thought the rump was perhaps darker than the rest of the animal, which had a tan colouring. He passed it and stopped the car but it had disappeared into the bush by the time he looked back. He also reported that a friend, Jan, told him that many years ago she had watched for ten minutes illuminated in the headlights, a pair of striped thylacines licking and preening each other on the roadside in the Snowy Mountains, in southern NSW

8th February 2006, Thursday morning around 5-30 -6.00 am, Mullumbimby "my 23 year old daughter Shanti saw what she believed to be a thylacine. We live on the eastern side of Mullumbimby town near the sugar cane fields at Morrison Avenue Mullumbimby. It was early morning around 5-30 - 6.00 am when she heard a commotion outside the house as if a dog was fighting with our cats. She went outside to investigate and rushed inside to tell me what she saw. She said she saw a "mutant" dog. It was small in size, light golden coloured with a very long snout and rounded ears, a long pointy tail, and stripes on its back.

It was making a strange gutteral yapping noise as it tried to attack our cats. It then chased one of them across the neighbour's garden. My daughter became frightened, as she said she thought it was a feral dog, and yet it did not look like any dog she had ever seen. She then saw it run as fast as it could down the street towards the cane fields. She said that its gait was awkward looking, and looked like it was loping because its front legs were shorter than its back legs, and it looked quite ungainly as it ran.

My other children and I heard the strange noise, but did not go out to investigate. The neighbour heard the commotion as well. My daughter did not know about thylacines before the sighting. It was only when my other daughter said that her description of the creature sounded like a thylacine that she was able to definitely identify it from pictures that she found on the internet.

She was able to get a good and clear look at it because it was so close.

15 February 2006, Hastings Point, Rose described a strange dog-like animal that she saw while driving to work in the morning. It was grey with dark grey mottlings on its rump, it had distinctive large round ears and was quite unlike a dog or fox. In December 2006 at 9.30-am she again saw the same animal with three young cubs chasing and playing together on the road.

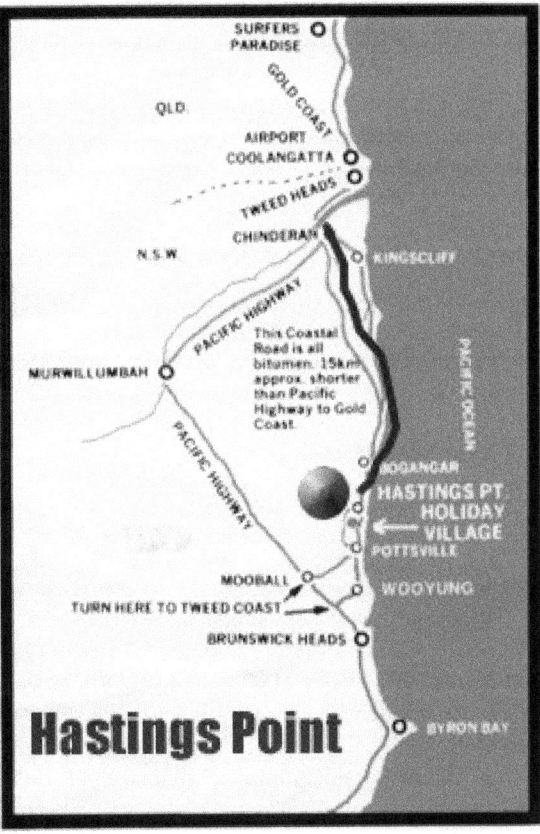

19 February 2006 at 5.30 am, North Ocean Shores. John & Pat had set off from their home at New Brighton for the Pottsville Market, where they would set up a pottery & jewellery stall, and had only reached the small bridge north of Redgate Road at North Ocean Shores travelling at 50 kph when they saw 2 strange animals standing on the road. Both were of a buff colour with distinctively rounded ears, hunched backs and remarkably long thin straight tails. They were somewhat smaller than a German shepard dog, much larger than a fox and one individual was smaller and was standing in front of the larger animal. The animals watched the approaching car for a few seconds and then raced off into the bush. Pam noticed what looked like pale stripes across the back. John was adamant that they were not foxes, dogs or dingos all of which he is very familiar with after spending 20 years at Lakes Entrance in eastern Victoria.

25th February 2006, Saturday night at about 11 pm, Palm Avenue in Mullumbimby. Kali saw a strange animal running across the road. She stated "No stripes, but a distinctly wild animal, reddish brown short hair, above knee-high. It really caught my attention, and I found myself thinking about this animal for days."

25th February 2006, Saturday night, Main Arm Road, Mullumbimby. Richard sighted a thylacine-like animal on the road as he drove from Main Arm to Mullumbimby and pulled off the road to look where the creature went.

26th February 2006, Mullumbimby, Left Bank Road, Sunday morning about 6:10 am. Alisha and her mother were going to the market and when they pulled out of Yankee Creek Road and went round the bend they saw a strange animal that was too big to be a cat and that was not a dog or a fox and it had stripes across the back, rounded ears & a long stiff tail.

26th February 2006, 3.15 pm, Redgate Road, South Golden Beach. Steve and Michelle and their family were riding their bikes when they saw a strange animal in the short grass. It was 60 to 70 cm high and covered with short ginger-blond short hair with a narrow, small triangular-shaped head, a long thin neck, a long straight, thin tail that was as long as the animal. It scratched itself with its hind leg like a dog. They did not believe that it was a dog, dingo or a fox.

February 2006, Brunswick Heads, on the Pacific Highway near the Mullumbimby turnoff, Walter saw a dead thylacine-like animal, with a distinctive long straight tail and stripes along the body, lying on the roadside, a victim of a car strike. I searched the locality but found no sign of a body.

February 2006, Shara Boulevard, North Ocean Shores Rayleen saw a thylacine-like animal crossing the road.

February 2006, Tyalgum. Donna lives at Tyalgum and saw an unusual animal that ran through their paddock on dusk, at a great speed. At first she thought that it was a wallaby but it's head was too large and it was too bulky in it's front end then it tapered off in the hind end and had a tail similar to a wallaby. Nobody in her family believed what she had seen until 3 days later when she was coming home with her daughter and about 2kms from home they saw the same animal run into the bush and her daughter described it exactly as Donna had seen it.

5th March 2006, Sunday, Left Bank Road, Mullumbimby. Elle and her family
saw a strange animal in their garden quite close to the house which ran down to the creek. She saw it a couple of times during the following days. She stated that "it looked funny and very
skinny and moved weird-like. I looked it up on the net and it did look like the pictures of thylacines but with no stripes. I thought it looked a silver colour but it was hard to tell because it
was raining."

11th April 2006, Shara Boulevard, North Ocean Shores 12.30 am. Ron was driving west towards the highway and saw what looked like a thylacine as it walked along the northern side of the road. He stopped the car & turned off the motor & watched it from 2 metres away in the high beam of the headlights as it stared at him. It was about 1 metre high and 2 & a half metres long with high haunches at the back of the body and a long thin pointed tail almost the length of the body & which pointed down. Its snout was narrow & pointed, the eyes sloping backwards, the ears tall & rounded and it had a long neck. It had very fluid movements & its body was covered in short hair of a light fawn colour. It had 10 to 12 very pale stripes across the rear portion of the body with slightly wider stripes on the back & narrower stripes on the rump. After looking at the Ron in the car it suddenly bolted off into the bush.

2nd November 2006 at about 9 pm, Thursday, New Brighton. Rob was fishing at Casons road New Brighton when they saw an unusual animal. He stated "it had a tail like a roo, but not touching the ground. I've shot and skinned roos and foxes before so I have had experience with them up in Armidale. We observed it for about 10 minutes from between 10 to 20 meters away. It walked, going from tree to tree looking up the trees. Its eyes were pale green reflecting in our weak head torches, the moon was in the clouds, but a bit of light was coming out. The animal on the cascade beer label would be the closest thing and its ears were not as big as a roo or a fox. It was grey/brown with a lighter underside especially under the chin. Its tail was darker and it had short hair just like a roo, it had no stripes. My friend who was with me describes it as "looked like a kangaroo but it walked rather than hopped". After shaking our heads in disbelief, we were the ones who left, it hung around in the distance."

January 2008, Byron Bay Andrew and his wife were on holidays from Victoria, staying at an apartment off Cemetery Road in Byron Bay. On a bike ride to the beach they had a good look at what both thought may have been a Tasmanian tiger "as ridiculous as that sounds" stated Andrew. "It had sandy coloured fur, a snubby muzzle and a long tail. It had the gait of a dog and certainly wasn't a feral cat. The stripes certainly were not bold. My wife works at the Werribee zoo in Melbourne and it certainly wasn't like any of the cats they have there. We then mentioned it in passing to the bus driver on the way back to the airport and he said there had been some sightings.

3 March 2008, Frazer drive, Tweed Heads
Joel, Frazer drive, Tweed Heads, near the Tweed Heights turn off, at about 11:40 pm on the 8/3/08

Hi I'm Joel from Tweed Heads area and I was travelling on Frazer drive in Tweed Heads, near the Tweed Heights turn off, at about 11:40 pm on the 8/3/08 and I saw this creature lopping across the road with its tail straight. I thought it was a cat at first but as I approached it, it didn't scurry as a cat would. I have looked at other posts and seen what they have described it as it matches it. Then, the next morning, 9/3/08 at approximately 6:40, as I was travelling to Murwillumbah on the Terranora road, about 1.5 km from the Bilambil turn off were it start to decline, I saw it again. Lopping once again & this time I was closer and still it made no sudden movement like it wasn't scared of my car. It looked, then lopped into the cliff side of the road. It had its tail out straight and it seems like the tail is very tense. Next time I see it I will take a photo. It has me very interested. I hope this helps & I would like to hear feed back about this case if that is possible?

Early 2008 – late 2007 Repentance Creek Rd at the Minyon Falls turnoff.

I, Zabloc, recently (early this year or late last) saw an animal near the Minyon Falls turnoff (Repentance Creek Rd) that could have been a Thylacine. It was a bit after dusk. As we were driving towards Rosebank, I saw something (about the size of a whippet - maybe smaller) dash across a grassy slope. It was travelling at an incredible speed (almost unbelievably fast). Earlier last year, on the road between Ballina and Eltham (just after the Alstonville turnoff), dad and my brother saw something similar: an animal of similar size passed in front of their car so close both thought it was certain to be hit. But it was so fast, it passed by unscathed. They both wouldn't have believed what they had seen unless the other had confirmed it. Both times it was too difficult to see it properly, because it was travelling so quickly. I'm not sure if this is a characteristic (speedy) of the Thylacine. I just thought you might be interested.

2003 Byron Bay from Ballina Road 2k's past the Lennox Head turnoff.

I, Doofgdaddy, have been loathe to write as my friends have a great laugh at my expense. About 5 yrs ago I was returning to Byron Bay from Ballina after work in the early hrs of the morning. I had large driving lights which clearly picked out an animal about 2k's past the Lennox Head turnoff. It was trotting quickly with a stiff legged gate. It had stripes and a thin tail stuck straight out. It stopped, looked in my direction then continued into the bush. It was not a dog. I thought a sick wild pig but it's build was just too thin and the tail was so long (and stripes?). Who knows?

2008 Mooball
On Sunday 22nd June at 3.00pm, Scott Green, the editor of the Weekly News, while out riding his bike along Wooyung Road towards Mooball, approaching the old railway bridge, spotted what he clearly thinks was a thylacine. It emerged from the vegetation 70m ahead of him to run across the road travelling north when it spotted Scott & stoped to watch his approach. Standing on a road that has only light traffic it stayed for about 40 seconds until Scott was within 40 m of it. It then ran into the vegetation.

He described it as having a head and body about 1 m long with a longish pointed tail about 60cm long. It looked like a stretched-out greyhound, 30% longer in the body than a dog. The back and the hind legs looked more like a cat, tan-coloured with orange tinge in the sunlight, stripes above and with a thicker coat than a dog.

2008 Upper Wilson's Creek
Tracey emailed the ABC North Coast Radio to describe her sighting
"We live in Upper Wilson's Creek and back onto Mt Jerusalem Nat Pk. We have seen this "mystery" animal at least twice now and in fact had assumed it was a wild dog and reported it to NPWS for dog baiting! Luckily, we were too busy to follow it up, but this animal is definitely at home on our 40 acres. Last close sighting was about 6 months ago and it was right by our car on our forest drive. We have been troubled by irresponsible dog owners whose pets have eaten our free range chooks (now all well fenced) so actually got out of the car to photograph the animal but it calmly loped off and it was too dark to get a shot.

It has a distinctly grey striped torso, long pointed muzzle, and a thick rigid tail. Spooky to see the drawing on the web site and realise what it might be!"

August 20th, 2008 Coast Road between Lennox Head and Suffolk Park
David Hall wrote at 11:39am on August 20th, 2008:

"I'm totally convinced I witnessed a Tassie tiger on the coast road between Lennox Head and Suffolk Park NSW. The animal was in the headlights of my car eatings roadkill in the middle of a road and then bolted into thick scrub. It seemed to be stationary for a second or two as it was down the end of a long stretch of road so I managed a pretty good look. I'm dead-set sure the damn thing was not a fox, dingo or dog and I'm positive about the zebra like stripes down the back. I wish I'd gone back to sit, wait and watch just to cure my curiosity. So from that experience I do believe they are still knocking around"

Russell's drawing of the strange animal that he observed around 6am on Australia Day, 26th January 2006, on Shara Boulevard, North Ocean Shores.

I'm working on the Brunswick Heads Bypass. One of the engineers here has seen this animal a few times and has managed to get a photo of it. He's actually seen it climbing a tree. You can't see in the photo but he says it's got stripes across it's rump.

Regards, Adam, 2006

FOLKLORIC CREATURES OF PHILADELPHIA AND THE IMMEDIATE COUNTIES
Neil Arnold

Philadelphia. The largest city in Pennsylvania, and the sixth most populous city in the United States. Yet, despite its legends of ghosts, and the fact that elsewhere in Pennsylvania is considered to be a haven for many cryptozoological creatures, especially Bigfoot and large, exotic felids such as black panthers, Philly, also known as 'The City Of Brotherly Love' is considered pretty much bereft of strange creatures. My research regarding the U.S.A. concerns looking at individual States and their levels of weirdness when it comes to 'monsters'. The likes of West Virginia, Ohio and Maryland have been covered time and time again whether in reference to hairy hominids, flying humanoids and all manner of oddities from Goatmen to white-furred humanoids, as well as classic UFO saga's, mystery Men In Black and countless ghostly tales. The fact that Philadelphia is ranked 49th as the world's most populous city may go a long way to explaining as to why there is no room for even the most elusive of beasts. Over 1.4 million people inhabit the city so if an unnatural, or shy creature awaiting discovery does prowl the shadows, surely it must've been found by now, or forever remain in the murky depths of folklore, where ghosts and other phantasmagoric manifestations and superstitions lie.

My exhaustive research however, in which I trawled the beliefs, fears and folklore of 'The City That Loves You Back', has in fact sifted out many extraordinary creatures from the metropolitan arena. I would also like to share with you a number of beastly legends from relevant neighbouring counties. Philadelphia is surrounded immediately by Delaware, Berks, Bucks, Montgomery, Northampton, Lehigh and Chester, giving this patch of counties the name of Freedom's Corner but counties such as Lancaster, Schuylkill (both Pennsylvania Dutch) and Monroe (Poconos I-80 East) are also monster territory, and counties which must not be ignored with relation to Philly strangeness.

Surely, if Bigfoot exists in the wild woods of Pennsylvania, and counties further a field such as Mercer and Fayette, then Philly must have its fair share of hairy, bipedal invaders ? Well, seemingly not. Whilst the heart of Philly may be bereft of sightings of the hairy humanoid we've come to know and

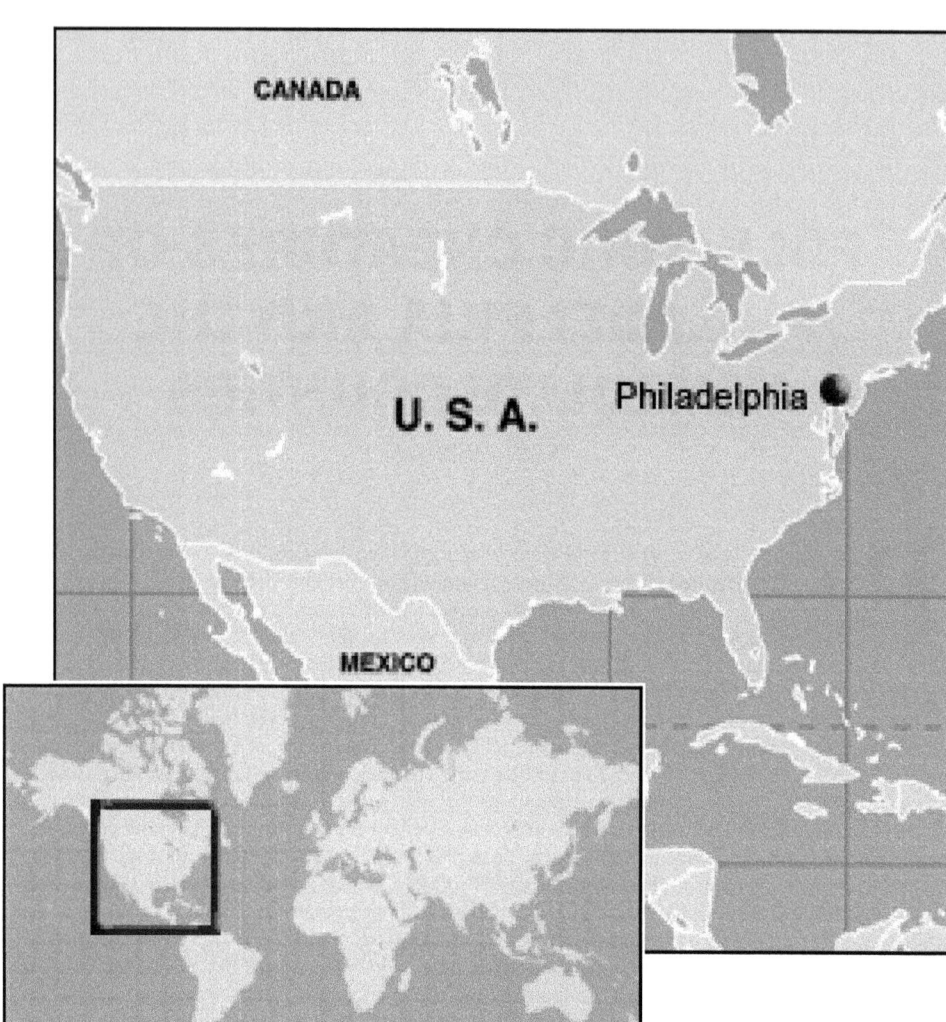

love as Bigfoot, or Sasquatch, history shows that such man-beasts have been evident only within the counties which surround.

During the summer of 1968 a sixteen-year old female camping in Bucks County was cycling to a site under construction when she encountered a creature, which stood just over five feet in height and resembled a monkey. The beast, which had two massive arms, was rocking from side to side as it leaned on a fence just eight-feet away. The woman encountered a similar entity some eleven years later in the same location. More campers at Bucks County saw Bigfoot during the '70s, with several sightings occurring in broad daylight. Another teenager who at the time was walking along a dried up river bed noticed a movement out of the corner of their eye and noticed a beige-coloured hairy creature crouching, that was still four-to-five-feet in height bent down. In 1976 two teenagers out hunting squirrels in Bucks County also saw a Bigfoot; the tall humanoid was around one-hundred feet away and casually chewing a tree branch. The creature was covered in reddish-brown hair and stood around seven feet tall.

Meanwhile, an extraordinary find - During the 1800s in the town of Sayre, Bradford County, Pennsylvania, a group of antiquarians made a startling discovery. W.K. Morehead, Dr. G.P. Donehoo and A.B. Skinner of Philadelphia's American Investigating Museum unearthed giant skeletons said to originate from 1200AD. However, the most significant detail of these seven-feet tall frames was not the size but the skulls, upon which were situated horns above the eye sockets! If such skeletons were freaks of nature then why did both skeletons bare similar protrusions?

At the time, as the remains were being transported to the Philadelphia Museum, concerns were raised that such remains should not be excavated, but instead left alone, as some superstitious folk believed the remnants were of demonic entities. Finally, when the over-sized humanoids were placed at the museum, they mysteriously vanished after a short time. The museum claimed they were stolen. Researchers stated that the horns were not mere bony lumps, which could have been a sign of disease, but horns measuring four inches in length. Were these elusive skulls merely a hoax, works of art, demonic remains or unknown humanoid?

Speaking of devilish things - The most famous legendary monster to sit so close to Philly is the Jersey Devil, a winged entity which has inhabited the New Jersey Pine Barrens for centuries, whilst remaining across the river. However, during the 'phenomenal week' of 1909 when the dragon-like critter was seen many, many times one particular encounter did in fact occur in Philadelphia.

The legend of the Jersey Devil was allegedly born around 1735, possibly offspring of Satan himself. A mutated alleged thirteenth child to a Mrs Leeds or Shrouds of Smithville, said to have been cast out in the blackness of night in Leeds Point, destined to roam the depths of the Pine Barrens for centuries to come. The legend has remained potent, sightings have continued over two hundred years after the so-called mythical birth, and some kind of creature has indeed roamed New Jersey. Many have described the creature as leather-winged, around four-feet tall, looking like a deer, but able to stand on two legs, having hooves and often heard screeching into the zenith. It has been hunted on numerous occasions, shot at, exorcised by priests, blamed for attacks on livestock, and yet it remains elusive. During that 'phenomenal week' of 16th January to the 22nd in 1909, a Mrs. J.E. White of Philly was hanging laundry at 4:00 pm when she observed a creature sitting in the corner, which, when she approached, rose to six-feet in height. It was covered in scaly skin and vomited flames from its mouth. When Mrs White's husband ran into the yard, the creature flew off. Seconds later a motorist claimed he almost hit a similar beast on the road and another witness, William Becker, reported he'd thrown stones at it.

Other sightings took place at Chester, and as recent as the 1970s in Mercer on the other side of Pennsylvania. However, the so-called Devil has always somehow eluded pursuers, embedding itself into monster lore. Bird, plane, Bigfoot, dragon or mutant? No one really knows, but it is better the Devil you don't know, than the one you do!

Among identities put forward for the legend of the Jersey Devil some researchers have suggested a hammered-headed fruit bat *(Hypsignathus monstrosus)* which inhabits Africa and referred to as a flying fox or mega-bat, however, despite its elongated and bizarre looking face, this creature certainly doesn't measure up to the size of the Jersey Devil and only eats fruit. Another suggestion is the Mountain Lion (Cougar, Puma) which have shrunk back into the most remote woodlands of Pennsylvania, or, as sceptics theorise, driven to extinction despite numerous sightings around the Philly area. In cryptozoological, and certainly folkloric circles, the Jersey Devil has changed its identity over the years, or indeed, been altered by society. At times the creature resembles Bigfoot or dragon to some witnesses, but at other times its shriek and predation of livestock points to a prowling large cat. The puma has certainly become enshrouded in a mythical aura as sightings have declined over the years. Pennsylvania's last pair of mountain lions were said to have been destroyed in 1871, yet over one hundred years later sightings continue throughout Philadelphia. Are witnesses merely seeing ordinary cats, or other local wildlife such as deer, and mistaking them for these slinking, fawn-coloured shadow cats? Or do the dark corners of the city hide a small population of creatures thought long gone?

Proof of such prowling animals is not exactly in abundance, hence the mythical status these cats have brought among themselves, but tracks, fresh kills on livestock, eerie screams during the night and countless eye-witness reports seemingly point to the existence of this predator which officials have dismissed for decades. However, during the mid-1990s a small cougar flap took place in Philly, particularly around Delaware and creeping into the suburbs of Philadelphia. Such animals, although denied, have vast territories of several hundred square miles and often live up to their reputation as being elusive, hence their nicknames 'ghost cat' and 'shadow cat.' However, one such cat was killed on the outskirts of Philly in 1967. Further proof of such cats hit the headlines in 1996 when an animal resembling a cougar was filmed and shown on Philadelphia area television. However, it's more than likely that the 'mythical' tag will follow such animals wherever they go, unless one unfortunate victim turns up dead for scientific analysis, and only then will it prove what we've already known for years, that the Eastern cougar is still alive and well.

It's a shame that the elusive cougar has been

thrown into the cauldron of local weirdness with other ingredients such as UFOs and ghosts.

There are even legends of huge snakes slithering close to Philly. In March 2008 it was reported that two unusual creatures had been seen in Bristol, PA, causing alarm among locals who live in the neighbourhood. The Delaware River town was seemingly overrun with large, out of place snakes, thought to be pythons. Police attempted to track down the beasts, commenting that one measured nine feet in length, and was an albino python, while the other was a twelve foot long animal, possibly a Burmese python, and that both might have escaped from a private collection or been released into the wild on purpose. After several thorough searches, police unearthed the albino critter, but the *Philadelphia Inquirer* believed that the snake still on the loose had been responsible for the disappearance of several kittens in the town situated a few miles upstream from Philadelphia. Locals were in a panic, claiming to have seen a snake very dark in colour that was rumoured to have found residence under a house in the neighbourhood, although searches proved fruitless. Cameras had been installed in the hope of tracking the monster.

A local man named Mr. Kile said he encountered the snake whilst searching his crawlspace for his missing cat, and then heard the hissing horror. It was spotted relaxing in a children's wading pool. Such reptiles can reach up to twenty-three feet in length and inhabit parts of Asia, and are also known as Indian pythons. Snakes are certainly one of the most commonly sighted exotic animals when it comes to appearing in places they shouldn't!

With regards to the US, it is believed that many such snakes established themselves in South Florida after an influx of collections which then outgrew their owners, who opted for the simple solution and

An American alligator and a Burmese python locked in struggle.
Photo by Lori Oberhofer, National Park Service.

released their pets into the wild. Bristol animal control officer Bill Kurko stated that this specific animal is not dangerous; such pythons will hiss and strike if threatened, but their bite is neither venomous or dangerous. So, the next time you visit the toilet....beware!

Despite a handful of stories pertaining to escaped pets and the possibility of roaming cougars, the tales of phantom animals and folkloric creatures are certainly those which appeal to locals. Is it because such yarns are merely the stuff of campfire legend, which, although having the ability to frighten, remain the stuff of chilly nightmare instead of woodland resident ? The black phantom dog is one of the world's most sinister and intriguing legends. Several theories have been put forward to explain such ghastly spectres which take on varying forms, but always have the characteristics of huge, fiery-eyed dogs also known as hellhounds. Such manifestations have been observed headless and often roaming old cemeteries, whilst on other occasions they have been seen wearing heavy chains around their necks. Ghostly hellhounds also appear red, white, purple, yellow and blue in colour, although black is usually their coloration, and they have been known to float, disappear into balls of flame, and often haunt lonely roads, sometimes in the vicinity of churchyards.

Such a beast is said to roam the area known as Darktown in Lehigh County, a place known for several other strange apparitions such as a troll-like figure often said to appear in the old, abandoned village. The place has strong connections to Native Americans but sits as a ruinous and eerie setting occupied by a large, spectral dog, mainly in an area known as The Alamo, situated near the Thomas Iron Works. Several spooky legends are tied to the location, and ghost stories have filled the night air for many decades.

In Delaware County, at the SPCA Cemetery, it's rumoured that an evil hound roams the burial ground. Such a beast, once known as Bear, is often seen in the vicinity of the graves where hundreds of pets and other animals have been buried in the past. The dog was said to have been a savage creature that once mauled a man and so was put down, but forever roams the foggy confines salivating through the graves. Those that have seen the dog claim it only appears after 11:00 pm, but such a grisly hound can be conjured should one be brave enough to cling onto a certain section of fencing. Those bold enough to hang on will surely see the burning eyes of the black critter as it emerges from the mist...Are such monsters omens of death, signs of impending doom? Do they exist merely as ghostly pets, or something more sinister, as guardians to the gates of Hell?

Often, the greatest mysteries of all are those never solved, and those which run for many years as mere whispers on the tongue, or as the chill that caresses the backs of a shivering group of campers huddled around a flickering fire, the orange glow lapping at their cheeks like the tongue of a woodland serpent - or like an innocent tree-branch.

There is a vague, foggy legend said to lurk in Lehigh County, mainly in the vicinity of the Bake Oven Knob Shelter, a log cabin like those featured in many a stalk 'n slash horror film, and said to be an ideal stop-off point for campers and the like who've spent many a weary day trudging the Appalachian Trail. No one knows whether it is a ghost that loiters in the Kittatinny area (there is legend of a spectral hiker said to walk the pathways) or some kind of escaped pet. It's appearances are accompanied by peculiar, eerie noises; many who have been terrified by the unknown visitor claim to hear murmurs and whispers, unsettling in nature. During the 1970s a small group of campers claimed to have observed a creature so incomprehensible and mind-boggling that it simply became known as 'it'. Descriptions, if you can call them that, suggested something nondescript, or maybe even non-existent, if that makes sense! Whatever it was that prowled close to the camp, it horrified the group so much that they left in the middle of the night. Despite the haziness, the legend is said to predate European settlers, and appears to have been mentioned in Indian lore.

So many urban legends and boogeymen from the hills, the skies, the darkest corners and the eeriest bridges, can take on any form, often determined by our own personal fears. For some, the beast of the Bake Oven Knob Shelter could well be Bigfoot, an elusive, red-eyed humanoid said to stand over seven-feet in height. Or it could be a misty apparition. Much of its substance depends on who frequents the area, and who fears it the most, and it's usually those campers huddled around the dancing flames who create such night terrors…

Ghost Mountain is another place of high strangeness. Situated in Bucks County, Buckwampum Hill is a place of history and old tales. Most of these are urban legends, from yarns pertaining to a race of cannibalistic albino psychos said to inhabit the dark corners, to spirits who rise from the Indian burial grounds that are housed deep within the rocks.

A bizarre monstrous spook said to be a large white humanoid, is often seen holding a baby in the vicinity of Fishtown's Palmer Cemetery. Considering the several thousand bodies harboured by the soil, wraiths are actually few and far between - but this particular ghoul has often freaked visitors out as it stares over the surrounding wall. At Hawk Mountain Sanctuary in Berks County, a ten-feet tall shining humanoid has been reported by many motorists who travel along the two-lane road to the peak of the mountain.

Although the legend of the werewolf spans across the world, real incidents of such encounters are either very scarce or simply never documented, but they do in fact exist. Lancaster County may be an hour's drive from Philly but I simply couldn't resist the following tale of wolfen weirdness to howl, or indeed shudder at! Wisconsin-based author Linda Godfrey, in her fantastic book, *Hunting The American Werewolf*, comments that: "With a Germanic heritage second to none, Pennsylvania has always enjoyed a lively folktale repertoire including werewolf, or Woolfmann, notions brought from the old country where trials of alleged witches and changesters remained vivid in emigrants' minds."
With that we venture to Lancaster, where sightings of Albatwitches, meaning 'apple snatchers,' have been recorded. Such humanoids were seen in February 2002. They were said to resemble stick-like figures, strolling by the road side. Such beings are covered in hair and stand around five to six feet in height, but are not bulky. Bigfoot researchers have argued that such creatures may be juvenile Sasquatch, but witnesses argue that the monsters they have seen have long muzzles and are not ape-like but very skinny. These spookily-named apparitions have been the belief and cultural dread of the Susquehannock Indians for many years, existing as local bogeymen. The Indians paid homage to these elusive monsters by carving images of these fiends on their battle shields, to create fear and trepidation. Despite having such a fearsome reputation, the Albatwitch is said to mainly feast on apples and is often considered a cunning creature which steals, usually from those who are relaxing with a woodland picnic, only to find their apples missing. The beast lurks in the shade of trees and snatches the apples, and that's where its eerie name originates, deriving from "apple snitch." It seems that the critters have ape-like behaviour if not looks and have been known to playfully throw the remains of the apples back at the people they stole them from.

The legend of the Albatwitch, which comes mainly from the area known as Chickie Rock, sounds like a vague yet unsettling spook tale. It is said that the monsters make their presence known by peculiar whip-crack noises which emanate from the dark woods, and although sightings still allegedly persist of these creatures it is believed they were driven to extinction during the early 1900s either due to lack of woodland or fading superstition, but the reality is, such mysterious forms resemble that faceless entity we all fear, that unnerving aspect of the unknown.

Surely it's time for a movie called The Albatwitch Project to emerge, detailing the misty fantasy of this long forgotten woodland prowler.

Similar dog-faced creatures, have been sighted in Wisconsin, where one particular entity became known as the Bray Road Beast, which Linda Godfrey has investigated for many years and another has been given the name The Michigan Dogman, a man-beast from Detroit.

Another eerie beast said to roam Lancaster County is the Goatman. Sightings date back to the 1970s when farmers recorded seeing a bipedal wolf-like animal stealing a chicken. The beast was grey in colour with a white mane, and had long claws, and on its head were two horns (which recalls the earlier mention of horned skulls discovered!). Meanwhile, Spook Wolves are odd entities said to roam near Philadelphia and throughout Pennsylvania. These beasts are said to be wolf spirits that were seen as far back as the 1800s. Legend has it that wolf corpses, stuffed and once on display at the Philadelphia Centennial, were in fact mere shells to house such evil spectres, and although such ghosts could never attack anyone they were often associated with the Devil, and seen by graves. Some researchers have suggested that such monsters are similar to phantom black dogs, or 'hellhounds,' which are also considered bad omens. However, it appears that many of these legendary creatures melt into one another. Bigfoot and Goatman have often been considered the same thing, as have giant hounds and wolfmen when seen running on all fours, so the question is not whether they exist, but how many different creatures exist?

Whilst on the subject of Goat-related things and spectral animals, I'd like to briefly mention a strange haunting in Philly at an undisclosed location. The house in question was always haunted. Shadowy figures, moving objects, peculiar noises and lights flickering. The family who resided there often encountered various strange phenomena, but seemed comfortable with it, until the black goat appeared to the kids, who were so shaken by the experience at first that they could not tell anyone. Later, the children of the house mentioned that whilst in the cellar they had seen an animal, a creature dark in colour, which they had chased around the basement almost playfully. This happened on several occasions, but the parents were of the opinion that surely it must have been a rat, or a cat belonging to a neighbour, but no, the children swore that what they'd pursued was in fact a goat-like creature, which always eluded them and scampered away into the shadows. Of course, like many ghostly tales, there is no resolution except to say that the family eventually moved on. Whether the spectre of the goat remained we'll never know, but it seems as though the creepy cellar was a place of limbo for the phantom and the children, after realising they'd been chasing a ghostly animal, were never to dwell in the dingy cellar again. Well, that's what the legend states...

In 1919 at Schuylkill County's Broad Top Mountain, legend persisted that a snake measuring more than forty feet had taken up residence. The creature was mentioned in Janet & Colin Bord's 1989 book *Modern Mysteries Of The World*, in which they stated: "There have been numerous sightings of giant snakes in America...although it is likely that a good proportion of them are escaped pets..." The creature was said to inhabit the coal-mine shafts during cold weather, and legend has it that the snake was first reported by hikers in the area and had been seen intermittently up until the '80s— but surely it wasn't the same snake?

A big snake was also said to loiter in the Lower Tumbling Run Dam in Schuylkill. This legend dates back to the 1830s, when the dam was built. Many fishermen inhabited the log cabins so as to fish for the local 'monster', a creature resembling a snake or serpent and said to measure around fifteen-feet in length. Of course, the elusive creature never made an appearance.

In the September of 1969, a strange creature was seen in Delaware and reported to police. Eye-witnesses saw a black animal standing no more than three-feet high with a two-foot long tail. The tail was tapered and curled. The strange thing about the animal was that it leaped and hopped, like a kangaroo or wallaby, and it left tracks measuring four inches long, but a search for the critter proved fruitless. It was suggested that witnesses had in fact seen a coati, a relative of the racoon from the Procyonidae family. These animals inhabit Central, South and South-West America and also Mexico. They have long snouts, can walk briefly on their hind legs, is able to climb trees and has a coat variation of red to yellowish, grey to brown fur. Across the US reports of what have become known as 'phantom kangaroos' irregularly reach researchers and the animals are rarely caught, like so many mystery beasts that make up so many of Philly's weirdest legends.

Finally, at Dauphin County, which joins Schuylkill, there is a waterhole known as Wolf Pond. The pond was once said to be inhabited by a huge fish resembling an over-sized pike. The monster fish was said to be black with gold bands across its back and also have a green head. The beast was mentioned in Charles Skinner's 19th century book, *Buried Treasure & Storied Waters, Cliffs & Mountains* and was said to be so big that it attempted to capsize a boat on one occasion, but the fisherman was brave enough to hit the immense fish with his oar.

Despite this listing of several monsters and mysterious creatures to roam Philly and the surrounding counties, it still remains one of the most inactive places in the U.S.A. regarding such mysteries. Let's hope the future brings more monsters out of the closet...

Sources

Janet & Colin Bord: *Modern Mysteries Of The World* (1989) Grafton Books
Charles Skinner: *Buried Treasure & Storied Waters, Cliffs & Mountains* (1896)
Linda Godfrey: *Hunting The American Werewolf* (2006) Trails Books
Philadelphia Inquirer, Phillyist, New Jersey Devil Hunters, Pennsylvania Haunts & History, UFO Info.

STRANGE ANIMALS IN SNOWDONIA
OLL LEWIS

If there is one place in the UK that still manages to retain an air of mystery into the 21st Century it is Snowdonia in North Wales. Walking through its picturesque valleys there are times when you could be forgiven for thinking dragons could be lurking round the next bend in the river, or a lake monster could be ready to break the mirror-like surface of a mountain lake. If some of the many legends told about the area are true, you might be right. Snowdon is rife with tales of strange beasts, from the bizarre water leaper to fearsome winged serpents.

The water leaper

The water leaper was said to have the body of a toad, but a tail instead of hind legs and wings instead of forelegs. Local belief in the water leaper was certainly strong enough in the 18th Century to prompt shepherds not to use their dogs around the banks of llyn Glas for fear of an over zealous dog herding an animal into the water where 'something' would catch hold of it, and pull the sheep to its doom.

John Rhys who was studying the folklore of the area in the late 18th Century transcribed the only detailed encounter with a water leaper ever recorded in print. The encounter took place sometime in the first few decades of the 18th Century; a local fisherman called Ifan Owen, also known as Han, had had an awful day's fishing.

Whenever Han had cast out that day something had nibbled at his bait and removed it cleanly from the hook without being snagged on it or pulling at the line. Because Han made his living from fishing he grew steadily more annoyed each time his bait was stolen and eventually, when he could take it no more, he moved to another spot, beside a small cliff in the valley.

When Han cast his line out here he felt something pull at his bait almost instantly and, not wanting to lose any more of his bait, he pulled his rod back much more sharply than usual to be sure of hooking the animal that had been getting away with his bait all day.

Having finally hooked his tormentor Han had to pull with all his might to get the creature out of the wa-

ter. After an epic struggle the monster erupted out of the water, and it shot off the hook towards the cliff, so fast that, according to Han: "*It dashed so against the cliff that it blazed like lightning*". Han later recounted that if it were not the Llamhigyn then it must have been the devil himself.

Han claimed that both he and his father before him had seen the water leaper on several occasions, and in a number of places along this stretch of water. It was said to scream loudly whenever a fisherman was able to pull it to the surface.[1]

Unlike many accounts of strange animals in folklore, this sighting had occurred within living memory of the time it was first committed to print. The person who related Han Owen's tale to John Rhys, who printed it in his 1901 book 'Celtic folklore, Welsh and Manx', was William Jones. He happened to be a descendant, on his mother's side, of the Pritchards, who ran the local pub and hotel in Bedd Gellert and invented the tale of Gellert - the martyred hound - to bring rich English tourists to the town. (See facing page)

The Pritchard family held regular tall-storytelling nights with relatives and friends from Bedd Gellert, the town in reality having gained its name for having been the gravesite of a missionary called Kellert in the 8th Century - not a brave dog - and the nearby parish of Dolwyddelen. Han Owen was regularly in demand for these nights as he had few equals in the area in his ability to spin a yarn, and it was at one of these nights that William Jones first heard Han Owen delivering the tales of his encounters with the water leaper. Whether or not the water leaper was only ever an invention of Han Owen nobody can say, but it has not been seen in the area since his tales faded from the memory of locals.

Gelert revisited

The Story of Gellert has become the most famous version of the martyred hound story. Prince Llewellyn the Great of Wales was given a puppy by Prince John (later King John) of England that Llewellyn called Gellert. As the years rolled by Gellert became Llewellyn's favourite hunting dog because of the animal's loyalty and the affection he showed him, so when it was time for Gellert to retire from the pack he was brought home to join Llewellyn's wife and baby son as a full member of the family. Gellert and the infant got along splendidly and seemed to form a close bond, and when the baby's mother died soon after, Gellert was always there to comfort the child.

One day Llewellyn went hunting to catch a dangerous wolf that had been taking farm animals, and even attacking people, and left his young son at home with Gellert to keep him company. During the hunt Llewellyn found the wolf near to his home and gave chase, but eventually lost it. This was an embarrassment for Llewellyn, who was famous for his hunting skills, and had promised the villagers that he would bring the wolf's rampages to an end, so he stayed out until the evening in a fruitless search for the wolf. When he returned home, Gellert ran up to his master with his tail wagging and his face covered in dried blood. Llewellyn rushed up to his son's room where his eyes met with a gruesome tableau. Blood was splattered all over the floor and up the walls, the cot was over-turned, tapestries, furs and ornaments were thrown asunder and a large amount of blood was pooled all around a suspicious looking pile of sheets.

Believing the dog had murdered the boy, Llewellyn ran his sword though the hound in a fit of rage. However, as Llewellyn drew out the sword he heard his son's cries from underneath the upturned cot. He ran over to the cot and found his son underneath it unharmed, so he ran over to the pile of sheets and pulled them up, to find out where the pool of blood had come from.

It was the wolf Llewellyn had been hunting. After it had escaped it had come to the house and tried to attack the boy, but Gelert had protected the child and killed the wolf after a violent battle. Llewellyn was distraught upon discovering his mistake and ran over to the dying hound. He held him in his arms and comforted him as he slowly bled to death.

As the loyal dog licked his master's hand while the last of its life ebbed away, the prince swore that he would build a memorial to the bravest and most loyal of all dogs atop the animal's grave so that nobody would ever forget what had happened.

Gwibers and the Welsh Dragon

Almost everywhere you go in Wales you'll hear stories about gwibers. The gwiber (pronounced 'why-bur') is the true form of the Welsh Dragon, a winged serpent. Descriptions of gwibers in welsh folklore differ enormously in size, ranging from about 30cm to several metres in length. Some were benign and others were said to have held whole towns in a grip of terror. According to folklore, the gwiber of Penmachno was one of the most fearsome dragons ever seen in Wales.

The story goes that a gwiber had been causing mayhem eating fish in the river Machno, raiding the farms of nearby Penmachno and killing anyone foolish enough to get in its way. So, in desperation, locals managed to raise a substantial reward for anyone who could rid them of the monster. This caught the attention of a brave and well-respected man in the area called Owen ApGruffydd.

Not wanting to go in unprepared, Owen decided to visit a local soothsayer called Rhys Ddewin, before he went after the monster in the hope that the soothsayer might tell him his future and, with it, how to kill the gwiber. Unfortunately for Owen, Rhys predicted that the monster would bite and kill him.

Understandably perturbed by this prediction, Owen went back home downcast. If the beast was to kill him then he couldn't have much time left in this world and whatever he might try to best the gwiber would be fruitless. In desperation, Owen resolved to test out just how accurate the soothsayer's predictions were.

Owen went to see Rhys Ddewin a second time, but this time he disguised himself as a tramp and gave a false name in order to fool the soothsayer. When he asked Rhys how he would die, the prediction this time was that he would die because of breaking his neck.

Smelling a rat Owen tested the soothsayer a third time. On this occasion he was disguised as a miller and Rhys gave yet another different prediction about the nature of Owen's death; that he would drown.

In fury, Owen tore off his disguise and challenged Rhys as to how he could die in three different ways. Unruffled Rhys just said *"Time will tell"* as Owen scoffed at him and went on his way to do battle with the gwiber. Owen felt that he had exposed Rhys Ddewin as a fraud and that he was not destined to die at all.

Buoyed by this false confidence Owen did not notice the gwiber's presence until it was too late. The winged snake pounced on him while he was atop a small cliff beside the River Machno, and sunk its fangs into his side.

Owen pulled himself out of the gwiber's jaws and in the struggle slipped on the pool of his own blood that was pouring out of his wound. He fell down the cliff breaking his neck in the process and landed in the River Machno, where he drowned unable to move his body to escape the current. His lifeless body washed up near the town of Penmachno a few days later.

Gwiber by Shiné

Gwiber by GT

For the locals, who had been fond of Owen, this was the last straw. Every man in the area was rounded up, and they took their bows, and any other weapon they could lay their hands on to form a mob to kill the gwiber before things got any worse. When they tracked down the gwiber it was sleeping on a ford in the river and the mob let fly with a hail of arrows all at once. The vanquished beast was washed down river and the valley was named the Wibernant valley in commemoration of the event.

There is another story about a gwiber from Snowdonia, that is often misquoted in English references as a wyvern due to the fact that both the word gwiber and wyvern come from the welsh word for viper. This gwiber was said to inhabit a lake in central Snowdonia called Llyn Cynwch. The Cynwch gwiber was different to the Machno gwiber because it could transfix animals by simply looking at them. When it had caught any animal in its gaze, including people, they would not be able to move and the gwiber would have to do no more than slither over to the helpless creature and swallow it whole. The gwiber's reign of terror was brought to an end when a shepherd chanced upon it sleeping off a recent meal. Seizing his chance to kill the beast he grabbed an

River Machno, near Penmachno, by Terry Hughs

axe and hacked at the serpent. After several blows the gwiber fell limp and died, but to be sure that this was not a ruse, and that the animal would not rise again, the shepherd covered the carcass with a pile of stones. The pile of stones was named the Carnedd y Wiber (Cairn of the Viper).[2]

Another tale often told in folklore featuring dragons was the prophecy of Merlin, or Myrddyn as he was called in the welsh language. Long before he met King Arthur, Merlin was known far and wide for his wisdom and gifts of prophecy. It was said that Merlin was born as the result of a union betwixt a demon and a human and the denizens of Hell planned for Merlin to be the anti-Christ. The plans of the demons were scuppered however when Merlin was baptised*.[3]

The story continues that King Vortigern was attempting to build a tower near to Nant Gwynant, but every evening a violent earth tremor would destroy the work that had already been completed. When Vortigern sought advice from his soothsayers he was told to find a boy who had never had a father and sacrifice him at the tower's base in order to quell the turbulent ground. Vortigern saw no problem with this so sent out a search party to find a boy that fit the bill, and they returned with the young Merlin. As Vortigern made preparations to sacrifice him, Merlin was able to convince the King that his death would make no difference to the success of the building project.

Merlin then told Vortigern that the reason the tower kept falling down was that two dragons were fighting in an underground pool beneath their feet. Vortigern's men dug deep beneath the towers foundations and found the lake. There they beheld two gigantic winged snakes, one red and one white. As they watched, the white dragon appeared to have the upper hand over the red one and looked like it would kill

Llyn cynwch by Gwylym James

it. When the red dragon appeared to be at it's lowest ebb it struck back at the white dragon and managed to vanquish it. The red dragon then flew off though the hole the men had made while the white dragon's corpse sank to the depths of the lake. Merlin explained that what the men had seen was a vision of the future: the red dragon was representative of the people of Britain and the white dragon represented the people who had started to invade the country; the Saxons. Eventually when it looked that all was lost for the original people of Britain there will rise from them a great warrior king who will unite the country and free the people from Saxon oppression.

Because of the vagueness of the prophecy it has been applied to several figures in folklore and history, the first obviously being King Arthur, who - it is said - united Britain and destroyed the Saxons so utterly in battle they did not attempt to return to Britain for a generation. Other figures include Owain Glyndwr and King Henry VII. All three are said to have used a red dragon as their symbol in battle, which attests to the age and power of this particular folk tale in Wales.

*Mike Mignola uses a similar story for the origins of the title character in his comic 'Hellboy', although Hellboy is put beyond the reach of the forces of evil by his liking for pancakes rather than a baptism.

Hoop snakes

Keeping with the subject of strange snakes J.A. Brooks uses a modern incident involving snakes in the introduction to his 1987 book *'Ghosts & Legends of Wales'*.

In 1955, not long after the Forestry Commission had started work on planting the Sitka Spruce stands of Gwydr Forest, near Betws-y-Coed, a foreman found his workers huddled together in fear. As the foreman questioned the workers they told him the reason. As they had been clearing bracken and other shrubs from the area so that the trees could be planted, several serpents burst out of the undergrowth. The snakes had careered past the men curled up like bicycle wheels. The men had followed these hoop snakes a short distance down the ride before the snakes disappeared into the undergrowth again. It was while the men were debating whether to continue working and risk encountering more of these snakes when the foreman showed up. The foreman thought this was a load of nonsense, called them Nancy-boys and ordered them back to work.

A few weeks later the serpents made another appearance in the same part of the Gwydr Forest. Two men from a different team of workers were walking along a ride when they saw a snake basking in the sun. As they approached the snake it curled up and wheeled away into the shrubbery and bracken alongside the track. After throwing stones into the bushes in an attempt to scare the unusual snake back out, so they could attempt to capture it, the men reported what they had seen to the Forest Office. Sightings of the hoop snakes continued among workers for the rest of the summer, but the snakes do not appear to have been seen in modern times.[4]

J.A. Brooks said that it was likely that these snakes had been released by a private collector into an area where they were unlikely to have much contact with people. On the face of it that does sound like a reasonable explanation. However, there's a problem with that; no known species of snake can curl itself around in a wheel and move in such a manner. Indeed if a snake could curl its body up in such a way its spine would be so different from all other snake species that it would have to be classified as another sort

of animal entirely. It is possible that the sightings if genuine, and not misinterpreted in their re-telling, could be of some unidentified species of skink, but even then no animal has ever been filmed, or observed by scientists, moving in this strange manner and it would seem to be a very inefficient way to move around. Another possible explanation of the sightings could be a misinterpretation of the workers' testimonies, it is thought that many supposed sightings of hoop snakes in the United States of America were sightings of sidewinders that had just been incorrectly recorded. It is also a possibility that the workers had seen hoop snakes in a *Pecos Bill* film or in a *Desperate Dan* comic strip and were just playing an elaborate prank on their superiors.

Other hoop snake myths include the Japanese cryptid/yokai Tsuchinoko and the legendary Greek serpent Ouroboros. Tsuchinoko is a

Gwydr Forest by Dot Potter

flattened snake that is also said to roll itself up like a bicycle wheel and has been depicted doing this in art dating as far back as the Japanese Jōmon Period (14,000BC- 300BC) and continues to be described by witnesses to the present day. Ouroboros was a symbolic animal that was often used in Greek art to symbolize nature's cyclical system of destruction and regeneration, and was commonly depicted as a snake eating its own tail in perpetuity.

Afanc

Another animal that, like the gwiber, crops up quite often in Welsh folklore is the afanc. Afanc are a type of lake monster that looks like a cross between an enormous fish with crocodile like skin and mighty jaws. It is likely, from this description and the fact that they often inhabit fairly cold large mountain lakes, that afanc are in fact enormous pike. From a distance a pike basking at the surface of the water does look to have a body shape similar to a basking crocodile and the pike's mottled dark green colouring will only add to this visual impression. Much larger than average pike have been caught in a number of lakes that are reputed to have been the home of afanc for hundreds of years, the most famous of these lakes being Llangorse Lake in the Brecon Beacons. In 1846 the largest pike ever caught in the British Isles was landed there, weighing 68lbs, more than double the average size of pike caught in Britain. Llangorse Lake itself has been subject to reports of afanc since the dark ages. The kings of Brechiniog

even had an afanc as one of their royal symbols. There has been at least one person savaged by a pike over 6ft in length in the last 10 years, reliable witness reports of giant pike and a photograph taken of a dead pike that would have dwarfed the 1846 catch.[5]

Afanc have, however, suffered from an identity crisis in modern times. Some people have confused afanc with beavers and others with dwarfs due to poor translations from the original Welsh descriptions. This problem crops up time and time again and is not helped by afanc having at least six different spellings (afanc, adanc, addane, avanc, abhac, abac) due to the fact that Welsh spellings, unlike English, were not standardised until fairly recently and even now retain two distinct dialects.

Snowdonia is home to one of the most famous afanc in Wales; the creature known as Teggie. Teggie first made a splash when (s)he was first spotted in Llyn Tegid (aka Bala Lake) in the 1920s and has been described by several witnesses ever since. Usually the monster is said to be about eight foot in length and resemble a crocodile, but with smooth shiny skin. This description is the one given by Dewi Bowen, who was the lake warden from the 1980s until recently, several fishermen and divers.[6] However, some sightings are more outlandish bearing a resemblance to a seal or the ludicrous 'plesiosaur' interpretation of the Loch Ness Monster so beloved by certain sections of the media.

The 'plesiosaur' sightings only started after the media decided that Teggie was 'the Welsh Loch Ness Monster' and probably started as a result of the suggestibility of witnesses who had seen newspaper articles of Teggie accompanied by pictures of plesiosaurs. Hoax sightings of Teggie are very common because the monster very rarely makes an appearance, possibly because if Teggie is a large fish the animal would not have to come to the surface, and people who come to the lake hoping to see a monster leave frustrated.

Because of the large amount of definite hoax sightings more doubt is often cast on the existence of Teggie than most other lake monsters, there is however one unique animal inhabitant the lake whose existence is certainly not in question. This animal is a species of fish found nowhere else in the world called the gwyniad (*Coregonus pennantii*). The gwyniad is a species from the salmon family that became isolated in the lake after the last ice age, however because of recent concerns about pollution in the lake and introduced fish like ruffe (*Gymnocephalus cernuus*) eating gwyniad eggs the species' continued survival was in doubt. Recently though the environment agency have been rearing captive gwyniad from the lake to boost the fish's numbers and plan to intoduce the fish to other lakes.

Teggie is not the only avanc reputed to have lurked in the waters of Llyn Tegid, an archaic name of the lake was Llyn Llion and it was said to have been here that Welsh folk hero Hu Gadarn did battle with a mighty afanc. The tales of Hu Gadarn are set way back in the mists of time when a people called the 'Cymry' arrived in Britain for the first time. The Cymry had supposedly been the first people to arrive in Britain and had come from another country in sunnier climes and were led by a man named Hu Gadarn known for his wisdom, bravery and strength. The story goes that not long after the Cymry settled around the banks of Llyn Llion (Llyn Tegid) the afanc that lived in the lake started to cause flooding in their farms and settlements. The settlers tried to kill the beast, but their weapons could not even pierce its hide, so they asked Hu to help.

Hu decided that rather than kill the beast it would be better to transport it to another lake where it could do no harm to the settlers. The lake he decided to move the creature to was Llyn y Ffynnon Las, which is the archaic name for Llyn Glas, said to be the haunt of the water leaper in the 19th Century. In order to do this the afanc was tempted into the shallows of the water by a girl it had grown close to and Hu waited until the creature fell asleep near her lap before wrapping chains around it. When the afanc awoke it tried to pull away, but was stopped by the chains that were tied to Hu's bannog oxen, animals similar to aurochs. Enraged by this, the afanc bit

The middle of the three fish is a gwyniad

off the girl's breast. The oxen dragged the afanc though the mountains to its new home, where it has been blamed for pulling sheep and birds to watery graves ever since.[7, 8]

An interesting story but completely untrue. The tales of Hu Gadarn were invented by the modern day bard Iolo Morganwg (aka Ned of Glamorgan) when he was doped up to the eyeballs on laudanum. Iolo was born in 1747 and was a genius forger; among his most audacious achievements were reveling 'previously undiscovered' poems by Dafydd ap Gwilym from the 14th Century. These were, of course, forgeries made by Iolo but were not discovered to be fakes until almost 100 years after Iolo's death.[9] Iolo based the tale on several small legends in the North Wales area at the time including another afanc legend involving King Arthur.

Carn March by Arthur Barfog

King Arthur was called by local people to sort out an afanc that was causing havoc in Llyn Barfog (aka the Bearded Pool) near Machynlleth. The mighty Welsh king didn't need a woman to draw the afanc into the shallows, as Iolo later wrote that Gadarn had, and jumped straight into the water with a chain in one hand. Arthur wrestled with the beast in the

Home of an Afanc?
Lake Bala pictures
by CFZ Operative
Lisa Dowley

water and managed to get the chain wrapped around the afanc before swimming back to the bank. At the bank the king and his horse pulled the afanc out of the water using the chain. The force needed to pull the afanc from his watery lair is supposed to have been so great that one of the horse's hooves smashed a hole into the bedrock surrounding the lake that can still be seen to this day. Once Arthur had pulled the beast to land, it was powerless and he was able to kill it with ease.

Wolverines of Llandudno

The pleasant Victorian coastal resort town of Llandudno lies a stones-throw from the North-Western border of the Snowdonia National Park, it was here that residents and the local RSPCA (the Royal Society for the Prevention of Cruelty to Animals) noticed an increase in the numbers of cats going missing in 2004. At first the RSPCA thought that there may be a cat thief in the area stealing or relocating cats for an unknown purpose. However, the truth was soon to be revealed.

Around the same time, the RSPCA started to receive calls from locals about a large grey animal that was terrorising cats. Many of the callers said it looked like a dog, others said it looked like a small bear, several claimed it was a big cat and some said it looked like no animal they'd ever seen before. Soon the animal was identified as a wolverine and it was thought that the animal was hiding out in the local woods and coming out to look for food in the early hours of the morning.

One pensioner, Roy Chambers, saw the wolverine attack his cat, Bibby, and related his story to the

Llandudno by Lizzie

North Wales Daily Post:

"*It was about 4am and Bibby woke us up by bringing a mouse in. I took it outside but Bibby followed me.*

"*Then I saw it across the road - the size of an Alsatian, grey and with a long snout. It wasn't a dog, there's no doubt about that, but it definitely wasn't a big cat either.*

"*It went after the cat, so I ran out banging and shouting, and picked up a handful of stones to throw at it.*"

Despite further investigation by the RSPCA, who re-interviewed witnesses after Chambers' close encounter, the wolverine has still not been caught four years later.[10]

The big hairy man of Nant Gwynant

Villagers in Nant Gwynant have long told a story about how a cave in the valley came to be named. Long ago villagers and shepherds in the area were plagued by a thief that would break into their homesteads. They would awaken to find that their goats and cows had been milked, food had been stolen and the best sheep taken during the night. This went on for some years and every time anyone laid a trap for the thief it never took the bait and the finger of popular suspicion passed from ne'er-d'-well to ne'er-d'-well with each suspect's guilt eventually being disproved.

One day a shepherd was returning from the mountains later than usual and spotted something strange; a huge burley naked man covered from head to toe in red hair, so thick that it was almost like fur, was resting on a neighbouring hill. The shepherd suspected that this out of place and strangely hirsute giant might be the thief that was plaguing the village so the shepherd snuck past the man without being detected and ran back to the village as soon as he was out of sight. When he reached Nant Gwynant he rounded up all the available men and they hatched a hasty plot to catch the hairy giant. Because it would seem this plan involved running at the hairy man and shouting loudly whist brandishing makeshift weapons this plan was, not surprisingly, unsuccessful. The hairy man bounded off on all-fours leaping over obstacles with the skill and precision of a deer. A watch was kept on the area over the coming weeks to see if the hairy man would return, and he did a few days later. Because the previous plan had failed, the villagers decided to loose their dogs on the hairy man instead, however this also proved unsuccessful when the man bounded off with a hare-like speed.

The villagers despaired that they'd ever catch the man, as he was too fast for even their dogs to catch, and one man came up with the idea of consulting a magician. The magician told the villagers to find a red haired greyhound with out a single hair of a different colour and this would be able to catch the man. After much searching and bartering with local towns and villages the people of Nant Gwynant found a dog that fitted the bill and proudly took him home. When the villagers next saw the hairy man they were ready with the red greyhound and it was set loose to catch the hairy man. The hairy man escaped again by leaping down a small cliff.

After everything they tried to catch the hairy man had failed, the men of the village reluctantly gave up and resigned themselves to the fact that the thefts would continue. However, one woman was so angered by her frequent losses that she decided to stay up every night and hide herself in the front room of her farmhouse to wait for when the hairy man decided to pay a visit. Sure enough, after a few weeks, the hairy man paid a visit to the wrong house and the lady was waiting with a hatchet. She remained hidden until the man had squeezed his bulky frame halfway though the window before she struck the hairy man with her hatchet. The unexpected blow cleaved off the hairy man's hand in

one blow and he recoiled back out of the window before the woman could smite him with a further whack. The brave woman dashed out of her door, hatchet in hand ready to finish the man off but by the time she had gotten outside he had fled. When the village awoke the next day and the men learned what had happened they followed the trail of blood the hairy man had left behind to a cave beneath a local waterfall. As the big hairy man was never seen again it was assumed by the villagers that he had died in the cave, so the cave was named the cave of the hairy man.[11]

The lizards of Abersoch

Abersoch by Eric Jones

A small distance from the North Western boundary of the Snowdonia National Park, on the Lleyn Peninsula, lies the small town of Abersoch. For generations, locals have told tales about a unique resident; the cenaprugwirion.

The cenaprugwirion is a 30cm long muddy brown lizard with a bulbous orange sized head. The cenaprugwirion has a long tongue that can capture flies and other insects, independently movable eyes and a large dewlap under the chin.

The name of the creature can either be translated from Welsh as the scaly (thing) verified to be born of the heather (Cen = Scales, Ap = Son, Rug = Heather, Wirion = We Verified), although it has more commonly been translated as 'the daft fly catcher' in crypozoological literature, it appears that this translation, although a very plausible name for the creature, may be incorrect

The animal sounds superficially similar in appearance to a chameleon and the cenaprugwirion is believed to live in burrows, which - as some species of lizard can dig very deep burrows - could help a lizard more used to warmer climates survive the typically cold British winters.

Dr Karl Shuker has suggested that the cenaprugwirion may be out of place tuatara, which escaped from a local private collection founded when tuataras were more numerous in New Zealand and easier for herpetologists to get hold of. A tuatara could, in theory, survive in the British climate and if

NEW ZEALAND TUATERA (¼ nat. size).

the cenaprugwirion is a tuatara it could potentially be the last surviving population of a particular subspecies, having died out in its native land but survived, in relative secrecy, on the other side of the world… If that's not strange then perhaps nothing is.[12]

References

1) Celtic folklore Welsh and Manx
2) Janet and Colin Bord, "Llyn Cynwch, Gwynedd", Fortean Times Issue 77, October 1994

3) Ronan Coghlan, "The Encyclopedia of Arthurian Legends", 1991, Element books, Dorset, UK.
4) J.A. Brooks, "Ghosts & Legends of Wales", 1987, Jarrold Publishing, Norwich, UK.
5) Oll Lewis, "the Monster of Llangorse Lake", Animals & Men Issue 40, CFZ press, Devon, UK, 2007
6) Janet and Colin Bord, "Ancient Mysteries of Britain", Grafton Books, London, UK 1986
7) Iolo Morganwg "Triad 97, The Triads of Britain", Y Myvyrian Archaiology, 1807
8) W. Jenkyn Thomas "The Welsh Fairy-Book", T.F. Unwin, 1907
9) http://www.iolomorganwg.wales.ac.uk/bywyd-ffugiwr.php (retrieved 1/10/2008 (exerts from: Geraint H. Jenkins "A Rattleskull Genius: The Many Faces of Iolo Morganwg" University of Wales Press, 2005))
10) "Natural born killer", The North Wales Daily Post, 11/06/2004
11) Rev. D. Parry-Jones "Welsh Legends and Fairy Lore" B.T. Batsford Ltd. London, 1953
12) Dr Karl P.N. Shuker "From Flying Toads To Snakes With Wings", Galde Press, Lakeville, MN, USA, 1997.

THE BLACK TAPIR OF BREVET - A MALAYAN MYSTERY BEAST RESURRECTED

Dr Karl P.N. Shuker

The most startling tapir news for quite some time emerged earlier this year, when Dutch zoologist Dr Marc van Roosmalen formally described a new, fifth species of living tapir - the South American dwarf lowland tapir *Tapirus pygmaeus*. In stark contrast, the unexpected rediscovery of a no-less-notable cryptozoological tapir just a few years earlier generated scarcely a mention from any but the most devoted tapir aficionados. Indeed, until now, this very interesting resurrection has not been documented fully (if at all?) in any cryptozoological publication. The following account is an expanded, updated version of my original coverage of this subject that first appeared as part of a multi-part 1990s *Strange Magazine* article of mine on cryptozoological ungulates, and which later resurfaced in my book *Mysteries of Planet Earth* (1999).

Largest of the five living tapir species, and the only one that is native to the Old World, the Malayan tapir *Tapirus indicus* is further distinguished by its striking 'saddle' of white, encompassing much of its torso and haunches; its four New World relatives are all uniformly dark. Naturally, therefore, zoologists were nonplussed when one of the adult Malayan tapirs sent to Rotterdam Zoo in spring 1924 from Sumatra proved to be entirely black, with no saddle.

According to a subsequent paper concerning this singular animal by Dr K. Kuiper of Rotterdam Zoo (*Proceedings of the Zoological Society of London*, July 1926), there were no previous records of all-black Malayan tapirs, and not even Captain K. Brevet (of the Royal Dutch-Indian Army), from whom the tapirs had been received, had ever heard tell of such creatures before. Notwithstanding this, the Rotterdam specimen, a male, confirmed that at least one could (and did) exist. Moreover, when, just a few months later, Brevet sent two juvenile Malayan tapirs to the zoo, one of these matured into a second all-black individual (except for a small grey stripe on the median line of its belly).

Yet although they were evidently Malayan tapirs, what was their precise identity, taxonomically speaking? Both specimens had been captured within the Babat district in the low-lying plains of Palembang, a district where the familiar white-backed form also exists - thereby eliminating any possibility that they constituted a morphologically-distinctive geographical subspecies. And as these two mystifying individuals were both from this same region, with no reports of any all-black tapirs elsewhere, Kuiper also deemed it unlikely that they were merely the product of a simple genetic mutation - i.e. a melanistic (all-black) morph that could appear anywhere and at any time within any population of white-backed specimens (like black panthers within populations of spotted leopards).

Accordingly, Kuiper looked upon them as representatives of a newly-emerging variety, not replacing the white-backed version in any specific area (and hence not a subspecies), but nonetheless possessing a specific geographical distribution. In July 1926, within his *PZSL* paper, he formally christened his newly-categorised variety *Tapirus indicus* var. *brevetianus*, in honour of its discoverer.

I ended my original *Strange Magazine* and book accounts of Brevet's black tapir by stating:

> However, it now seems much more plausible that this all-black form was nothing more than a melanistic mutant after all, because no further *brevetianus* specimens have ever been documented. And both of the Rotterdam individuals died before any matings with white-backed specimens could take place - thereby denying science the opportunity of investigating the genetic basis of their uniformly dark colouration. Even so, their preserved remains at the Leiden Museum of Natural History bear silent witness to their erstwhile existence, and to the tantalising prospect that at some stage in the future, their kind will reappear, reviving the *brevetianus* zoogeographical paradox - the presence of all-black tapirs in the Old World.

Previous page: The only two pictures known of this rare variety of Malayan tapir.
Above: Artist's reconstruction

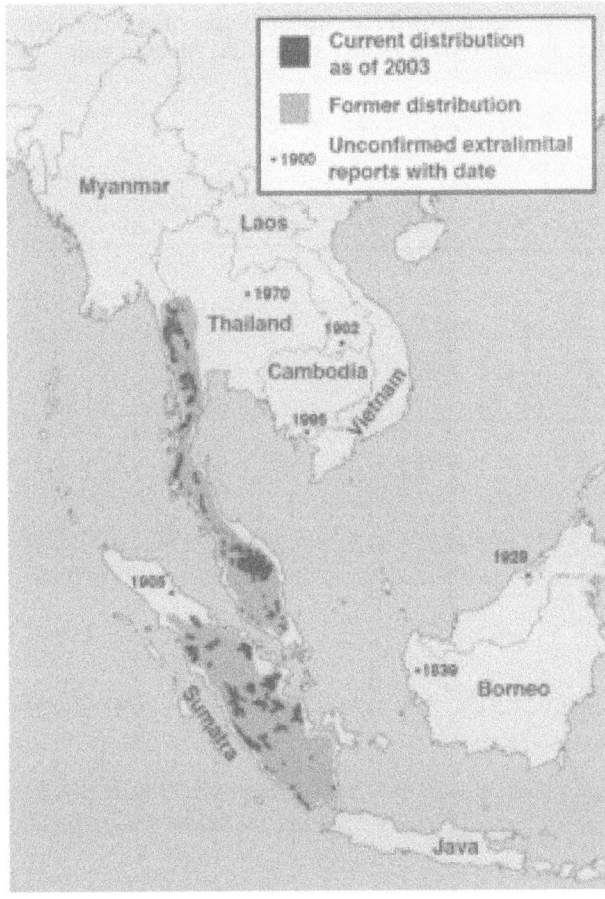

And indeed, after almost 80 years, that hypothetical stage in the future finally became the very real present, when in 2000, as subsequently documented by Mohammed Azlan J. of WWF Malaysia (*Tapir Conservation*, vol. 11, no. 1, June 2002), two separate all-black Malayan tapir sightings were recorded within Jerangau Forest Reserve, in Ulu Terengganu, peninsular Malaysia.

The first sighting took place on 9 July 2000 at 7.44 pm, when a motion-sensitive infra-red camera set up in lowland forest to monitor tigers obtained a clear photograph of a tapir that was wholly black, with no white saddle marking whatsoever. The second sighting, again the result of triggering an infra-red camera but this time set up in hill forest, snapped an all-black tapir at 1.13 am on 20 July 2001.

These are the first, and currently the only, photos obtained in the wild of specimens of Brevet's black tapir. Indeed, it might even be that both photos are of the same single animal, which may simply have moved from low forest into hill forest during the intervening year between the snapping of the two photos. Also, according to Wikipedia's entry for the Malayan tapir, in September 2003 Canadian researcher William Sommers witnessed the live birth of an all-black Malayan tapir in the wild. If correct, this is particularly intriguing, because tapirs of all species are normally born striped, only losing their markings as they mature (as was the case with the second all-black Malayan tapir at Rotterdam Zoo).

In any event, Azlan's photos provide conclusive evidence that Brevet's all-but-forgotten melanistic mystery beast still exists - albeit most probably as an exceedingly rare mutant limited to a couple or so individuals in every generation. And its unexpected appearance in Malaysia, greatly expanding this form's known distribution, substantiates my belief that it is indeed a mutant morph capable of appearing anywhere within the total Malayan tapir population. Benefiting from modern-day advances in DNA analyses, and with at least one contemporary specimen in existence, it would be fascinating to investigate the genetic make-up of Brevet's black tapir, and finally establish after many decades of scientific obscurity the true identity of this enigmatic creature.

Incidentally, in the belief that such a creature would prove exceedingly popular as a pet in the USA, biological engineers have recently attempted to create a dwarf version of the Malayan tapir – but that, as they say, is another story...

* * * * * * * * * *

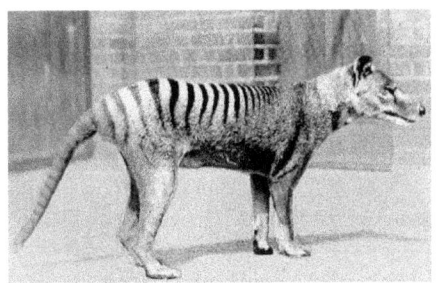

SPIRITUALISM AND UFO MYTHOLOGY
Dr David Waldron

Editors Note: whilst this article is specifically about UFO mythology, and was given to us purely for our own interest after the author and his father visited us this last summer, it holds - I believe -enough object lessons for forteana as a whole, and cryptozoology in particular, to warrant its inclusion in this current volume.

In his article 'The Psychology of UFO Phenomena', Saliba argues, that the public interest in UFOs and the concomitant phenomena of claims to abduction experience is well documented by the vast array of representations in popular culture, the many publications on the subject and also through quantitative analysis such as surveys and census data. Furthermore, Saliba makes the point that this interest is not only linked to the speculative efforts of science fiction literature and cinema but is also manifested in the explosion of scientific knowledge about the universe. Three Gallup poll surveys investigated by Saliba found that approximately 50% of the US population believed in the existence of extra-terrestrial beings and while approximately 10-12% of the population claimed to have seen a UFO, close to half the respondents believed in extra-terrestrial contact with people on Earth as an existent reality. This widespread belief and acceptance of UFO mythology raises the questions of the semiotic and symbolic representation in popular culture and the issue of whether scientific studies into the empirical veracity of the phenomenon is the most appropriate methodology of investigation of UFO and abductee/contactee experiences.

Typically, the debate surrounding popular beliefs and experiences of UFOs has fallen into two camps. Firstly there is the skeptic/believer debate surrounding the empirical veracity of UFO experience. From the believer perspective there is an attempt to define UFO sightings, mythology and contact/abductee experiences as an existent reality and thus eschew psychological interpretations in favor of either the language of scientism or that of transformative religious experience often phrased in a millenarian or apocalyptic worldview. Alternatively the skeptical approach seeks to debunk the experiences and beliefs of UFO believers and abductee/contactees and reconstruct them within the debate of scientific rationalism. Typically this revolves around presenting alternative models for interpreting these experiences through psycho-social phenomena such as hypnotic suggestibility, fantasy proneness and misperception. This approach to the experience tends to lend itself towards a form of dualism between two opposing camps of enlightenment rationality and science on the one hand locked in battle with a perceived superstitious credulity of believers on the other.

This dualist model of the conflict between skeptic and believer is quite disingenuous in that it orients the debate towards an empirical analysis that ignores the social, cultural and anthropological factors that shape, and perhaps construct, interpretations of UFO associated experiences. Furthermore, if the experiences are primarily apprehended as psycho-social in nature, this constructs the experiences as a symptom or illness to be treated as a psychological problem. Indeed it is argued by Lynn that psychological theories treating UFO and Alien sightings as a psychological problem, in isolation, are inadequate to explain these phenomena. This is particularly relevant due to the inherently embedded nature of the experience within the broader context of religious discourse and popular culture. In this context UFO related experiences originate as much in societal and cultural forces as in psychological and empirical experience and, furthermore, which is compounded by the hermeneutic impact of the therapist's own perception and belief structure surrounding the abduction/UFO sighting phenomenon.

Carl Jung articulated this dilemma when he argued that the debate surrounding the existence and significance of UFO related beliefs and experiences has been misconceived as an argument regarding whether UFOs are real or psychological in origin. The issue for Jung is that either UFOs are an existent empirical phenomenon which becomes profoundly psycho-social, symbolic and mythological in conception and experience or that they are entirely a psycho-social manifestation and entirely myth and symbolic forms. The underlying contention is that even if UFO sightings and abductee/contactee experiences are real - and it is worth noting that applying a unifying source to such a diverse range of phenomenon would be a major unsubstantiated claim - the layers of psychological, symbolic, religious and mythic significance ascribed to these phenomena indicate an endemic pattern of psycho-social significance. This issue is exacerbated by the intensely religious discourse used by abductees/contactees to describe perceived UFO and alien related phenomena. This is particularly pertinent when one examines the links between these experiences and the multitude of UFO oriented religious movements and the multi-layered and immensely complex UFO mythology manifested in popular culture and in UFO oriented sub-cultures and communities. As Jung argues,

> Such an object [UFO] provokes, like nothing else, conscious and unconscious fantasies, the former giving rise to speculative conjectures and pure fabrications, and the latter supplying the mythological background inseparable from these provocative observations. Thus arose a situation in which, with the best will in the world, one often did not know and could not discover whether a primary perception was followed by a phantasm or, conversely, a primary fantasy originating in the unconscious invaded the conscious mind with dreams and visions... In the first case an objectively real, physical process forms the basis for an accompanying myth; in the second case an archetype creates the corresponding vision.

From Jung's perspective it is a misconstrual of the psychic nature of the experience to argue that if UFOs and related phenomena are empirically verifiable then people have specific experiences, interpretations and construct UFO mythologies simply because they are existant. Similarly, to argue that if UFO's are not empirical phenomena then the vast array of experiences, mythologies and cultural formations linked to alien mythologies are simply mis-impression, hallucination or neurosis, ignores the profound, complex network of psycho-social experiences and structures that give meaning and insight into contemporary societal, cultural and religious experiences. The person involved in these experiences is far from a passive observer and the impact and significance of these events are inherently symbolic and cultural. More to the point, these experiences form the basis of, and are shaped by, a complicated web of social, cultural, symbolic and anthropological factors and are experienced in the context of a society, community, a mythological and religious structure and popular culture as manifested through decades of literature and cinema. It is a profoundly organic structure by which experiences both shape and are shaped by a web of inter-related social formations, structures and phenomena. In this light the rise of UFO experiences, religious movements and their immense impact in literature, popular culture and mythology and cinema yields significant insights into the corresponding relationship between popular culture, religious experience, pop mythology and cultural forms.

However, despite the insights the phenomena can yield when viewed anthropologically the study of them is hampered by two factors. On the one hand is the phenomenon's association with a stigmatized fringe and religious endeavor and on the other, the long standing and conflicting attitudes of the public towards these kinds of experiences and beliefs. In particular, the tantalizing but inconclusive nature of the physical evidence in this field (e.g. chemical changes in the soil and vegetation, electromagnetic disturbances, radar returns, photographs) has tended to derail scientific attempts to study the phenomenon and not a great deal of research has been directed to its historical, cultural and anthropological significance. Ufologists have long lamented that, for the most part, all they have to go on at base are witness reports and the activities of fringe communities and religious groups. Indeed, perhaps the most central theme that emerges in the study of UFO phenomena is the extraordinary ephemerality of the experiences and the proposed evidence. As Jodi Dean argues in her ethnography of UFO religious movements,

> The experience seems to have happened and not happened, takes place (or doesn't) in dream, everyday reality, visions and some kind of other dimension and resists earthly temporality. The experience seems, in other words, to be like a contemporary entry into the timeless world of myth, leads us to the intrinsically trickster cosmos of UFOism and the shamanistic nature of many such encounters and voyages.

Indeed, as Jung argues in his own analysis of the phenomenon, "the conclusion is that something is seen, but one does not know what. It is difficult, if not impossible, to form any correct idea of these objects, because they behave not like bodies but like weightless thoughts."

In this sense perhaps the most productive approach to investigating the significance of Alien Abduction mythology is to take apart its import culturally and socially in terms of the ethnography of the movements and the structure of significations surrounding the symbolism associated with UFO beliefs. Indeed tracing the history of UFO beliefs and the process of evolution by which the current matrix of popular culture and spiritual experience have been developed is something that can be studied with some surety of detail, unlike the empirical veracity of the experience itself. In Dean's research into UFO religious movements (and this is paralleled by research into abductee/contactee experiences) there are several common elements which seem to unite the experiences, beyond the many differences in narrative, symbols and visual representation.

To begin with, the UFO phenomenon is not confined to any one society. However, the exact manner in which UFO oriented cultural forms relate to the modern global economic system and the cultural influence of cultural exporting nations (such as the US) is another question and one which deserves serious investigation. Most studies tend to be limited in cultural and geographical scope and are often dominated by an Anglophone and US-centric approach to the phenomenon and beliefs surrounding it. One particularly pertinent issue in this respect is how much this Anglophone and US centric approach to UFO beliefs and experiences renders collective approaches to the issue problematic. As noted theosophist, Ian Blake, remarked in relation to such experiences,

Having personally experienced this odd sensation on two separate occasions I am reluctant to dismiss it as the subjective reaction of a high strung temperament. On the other hand, however, I am equally reluctant to interpret it as some form of rapport with extra-terrestrial entities. I suspect that most investigators would share my reluctance. There is a tendency nowadays, particularly amongst parapsychologists in the UK, to dismiss the Extra-terrestrial hypothesis as little more than a form of American cultural imperialism, rather on par with Coca-Cola, McDonalds and Ninja Turtles. It is far more likely that we are dealing with some form of psychic response, the precise nature of which is at present a mystery.

Furthermore, UFO reports and abductee experiences vary significantly between and within societies. In particular the overwhelming tendency to group together disparate forms under the heading of UFOs or Alien encounters obfuscates more than it enlightens. Ufological and orthodox scientific investigations have tended to look for patterns in reports and cast off variations even though these variations are integrally part of the experience. This is another area where an ethnographic approach can yield a great deal of information, not least of which by the examining the UFO experiences and folklore from within the social and cultural context in which they occur. The means by which the symbolic role of the Alien is configured as the quintessential expression of otherness, gives profound insight into social and cultural anxieties and social structure in the contemporary world.

In this light, the mythology and folklore of UFO experiences are not static but constantly in a state of flux and development. The images and imagery associated with extra terrestrial experiences have shifted rapidly. The shapes and body types associated with the Alien, and the ships they are believed to travel in, have changed remarkably in an extremely short period of time, within living memory in fact, yet strangely this seems to have escaped both believer and skeptic alike. A case in point is that stories of alien appearances in the 1950's and 1960's were of small slight green beings with large pointed ears and antennae. Even a cursory examination of the literature reveals this. Yet from the 1980s onward we have the now archetypal alien with a slight body, hairless skin, large round eyes and small mouth. What is particularly significant about this is the fact that the original stories and reports circulating out of Roswell, for example, describe the first alien type in the 1950s and the second alien type in interpretations of those same reports even amongst the same people who were positing the earlier type themselves at the time of Roswell. Indeed if there is one constant theme to the UFO experience it is the constant process of unselfconscious change in stories, mythologies and appearance in the literature and the tendency to collectively group stories together through the lens of contemporary interpretations and folklore. This is also manifested in the experiences and eyewitness reports of abductees and contactees and is a phenomenon that continually recurs throughout UFO oriented ethnographies. There is a constant process of rapidly transforming experiences and personal narratives of events which are seemingly unselfconsciously accepted by the contactee. Furthermore, not only will a narrative undergo a process of constant change and evolution, but also one's life experience and construction of identity are retrospectively reviewed and reconstructed through the lens of the alien experience. This is a process well illustrated in Whitley Strieber's biography of Alien Abduction, "Communion".

> The more I thought about it, the less able I was to accept that this had been happening to me most of my life... If I accept that this has happened and that it was buried even more completely than the events of October 4, then what else must I accept? Inevitably, that my conscious life was nothing more than a disguise for another reality. It is easy to speculate about such a thing on an idle evening, but when one considered the terrific intensity of the experience I had remembered, thinking that this might have happened again and again had the potential to shatter me.

Another prominent theme that Dean's research elaborates upon is the shamanic nature of the abductee/contactee experience. Firstly, before having an Alien contact experience people would typically view similar experiences and beliefs to be best interpreted and explained scientifically. After the experience it would typically be interpreted in a religious or pseudo religious light. The experience would be described as "extraordinary, life changing" and would use "the language of awe or terror." The experience is intensely personalized: it is linked to a paradigmatic transformation of world view and the experience is perceived to be profoundly symbolic and spiritual rather than as a passive empirical experience. The Aliens themselves are attributed with dialogue concerning religious or theosophical belief, personal spiritual growth and often engage in apocalyptic discourse. There are commonly alchemical symbols upon the walls and outside of the spaceship describing images of personal growth and transformation. Furthermore, the dialogue is often constructed in the language of proclamation. As Dean comments in her ethnography,

> Caywood told me, 'What will be found now will be found then.' Recounting "information" that she remembers/received from her Alien encounters, Debbie Jordan intones, "You have been blinded by your lives. You have let your negativity and your fear keep your inner eye closed. Do not fear God is life, eternal. Our greater good is that which works together to bring to the Spirit that which belongs to it.

In this sense, the experience is typically viewed as personally and directly linked to one or another aspect of the contactee's psyche that they identify as a source of particular anxiety to them. For example, In Mack's psychological research into abductee experiences he describes how an abductee claimed an Alien was drawn to her because she was dealing with "overwhelming issues of anger, frustration and being judgmental". Another contactee repeatedly discussed how the Aliens seemed devoid of emotions and could not understand her innate femininity and spiritual connection to the Goddess. Another respondent reported on how she felt the Alien beings were wanting to feed and drain her mental energy that she needed to be able to function in social life and work in which she took a great deal of personal pride.

Another important aspect to this intensely personalized experience is the common experience of sexual violation amongst abductees. Alternatively a person may describe mental violation in intensely sexual terms. Typically this sexualized experience is described as both deeply personal and yet also mechanistic in application. Many respondents also articulate that mental and physical violation seem intricately and symbolically linked with the bringing forth of deep psychological truths that the contactee holds dear to themselves. As abductee Eva claims, "In one encounter I learned that the impregnation process I experienced served as a process for the conscious awareness into an individuated being who is on a path of self discovery back to God, learning and experiencing along the way wholeness, completion, nonattachment and, most of all, humility.

In this sense, the more one looks at narratives of Alien contact experiences the more it becomes apparent that the original experience is invariably ambiguous and ephemeral but serves as the catalyst for profound psychological and social change. In essence it is a spiritual experience which can be examined in terms of psycho-social and cultural impact and religious phenomenology and its cultural context rather than an objective physical manifestation. It is a prism through which one's world view is profoundly changed. As such it has a profoundly shamanic and even liminal quality in which one faces the quintessential representation of otherness and what that evokes in terms of a search for symbolic and spiritual truths and metaphysical representations of cosmology and sense of being. The common thread uniting these experiences is a sense of timelessness and deep penetration into the body and psyche of the individual as an integral part of the abductee experience. As an abductee interviewed by Jodi Dean argued,

> First, it should be noted that I do not (or at least work to not) recognize the next mo-

ment or the one after that or the one after that *ad finitum*... as the future. My new paradigm in direct relationship to my abductions have shattered my former notions about space and linear time. So much so, that I have given up on the whole business and concluded that linear time models such as future, past and present are only useful when dealing with mortgage payments, time clocks, bus schedules and while waiting for my next coffee break at work.

It is in this light the Mack and Dean argued that the abductee/contactee experience essentially forms a shamanic spiritual role in contemporary western society. It serves as a source of transformative spiritual experience. There is an apocalyptic doom or rebirth of the planet (whether it be aliens as ecological caretakers or violating invaders) and a yearning for a new age of spiritual enlightenment, as well as a sense of the ordered, the rational and the scientific being overwhelmed by the strange and the uncanny. Mack, in particular, also argues that there is a strong commonality between narratives of cosmological journey and spiritual beings in narratives of shamanism amongst tribal people and that of abductee experiences. This latter claim quite concerns Dean who, whilst arguing there is a strong religious and Shamanic component to abductee experiences, argues that the direct conflation of abductees with tribal shamanism represents an orientalising of the experience through a 'conflation of emotional experiences, spiritual claims and native traditions with little regard to history and much opposition to a unified Western Scientific perspective.' Indeed, Dean later claims Mack's conflation of native American Shamanism with Abductee phenomena increasingly represents 'A New Age eliding of a variety of different experience, different cultures and different histories under the sign of personal transformation via great emotional enthusiasm and an expression of concern for the planet.'

However, this conflation, whilst most certainly an example of orientalising abductee experiences through a direct association with tribalism, irrespective of socio-cultural context, history and cultural specificity, does bring to light the theosophical antecedents of the Alien mythos. This is particularly significant as it represents a manifestation of the theosophical and occult origins of many aspects of the UFO mythos through the abduction experience. Other scholars, notably Rothstein and Denzler, have also written on the close parallels of Alien religious movements and theosophical approaches to spiritual experience and cosmology. Additionally, many of the most prominent abductees, such as George Adamski and Francis King, were quite steeped in popular theosophy. A significant component of this theosophical orientation is linked to the difficulty in accommodating Alien belief and abduction narratives within the confines of Judeo-Christian religious frameworks. As Denzler argues, in the US in particular, Christian responses towards Alien Abduction narratives and UFO mythology tended, overall, to either dismiss the veracity or significance of the experience by following the skeptical model or equate UFO experiences with demonic activity. Perhaps the most drastic of the latter was Pat Robertson's call to have abductees stoned for consorting with demons.

However, there is more to this shamanic and theosophical orientation than simply a perceived hostility and skepticism on behalf of some Christian religious organizations. As previously discussed, there is an intrinsically spiritual experience of displacement of self and journeys into other realms intrinsic to the UFO experience which is closely linked to both theosophical models of transcendent religious experience and millenarian interpretations of human experience. One of the most significant of these is by the student of famed occultist Aliester Crowley: Kenneth Grant, who played a significant role in the creation of a linkage between the mythology of Alien contact with that of spiritualism, theosophy and the fictional work of H.P. Lovecraft. Certainly, the most prominent of his contributions to UFO folklore was the popularization of the 'Greys' as the archetypal representation of the Alien form the Little Green Men of the 1950s.

Grant, as well as a major prosletyzer of the UFO abductee and grey mythos, was also the successor, as head of the Typhonian Ordo Templi Orientalis (OTO), after Aleister Crowley. Grant was quite enthused by the import of extraterrestrialism and contactee/abductee experiences as a new form of spiritualism that linked both the rise of new forms of science and space exploration with the theosophical esotericism of the OTO. It also linked in closely with Grant's strong affinity with the Cthulu mythos of famed horror fictionalist H.P. Lovecraft. In particular, Grant felt that the rise of new sciences and space exploration transformed the nature of the traditional religious cosmological model of Earth, as an ordered homocentric universe created by a divine being, into a chaotic disordered universe in which humans are only a small part of a much larger whole. As he argues,

> Astronomy and physics had shown the greater predictive value of Kepler's model, which placed the sun at the center of a solar system in which planets moved with elliptical orbits. The sun itself had been relegated to the stature of a fairly average star among an unimaginably vast number, all separated by immense expanses of cold and empty space. Geology had shown the earth to be whole orders of magnitude older than permitted by Biblical accounts. Paleontology had established the existence of a long series of different species that appeared at intervals, rather than in a single creation as required by Genesis.

Increasingly, Grant's fascination with extra-terrestrialism became the core of post-Crowley Typhonian spiritualism. Grant, who had a long tradition of interest in the notion of extra terrestrial beings, particularly inspired in the aftermath of the highly publicized Roswell incident, claimed in 1955 that he had discovered a trans-plutonian planet called Isis and founded a New Isis Lodge of the OTO for the express purpose of contacting extra terrestrial beings. He claimed that Lam, the archetypal "Grey" alien of contemporary UFO mythology, represented an Alien entity from this trans-plutonian planet which offered a way into a higher plane of being. Several groups of other theosophists and occult groups claimed to make contact throughout the 1960s and 70s, most notably Michael Bertieux.

In part inspired by the work of Eric Von Daniken, the fiction of H.P. Lovecraft and an extra-terrestrialism of Madam Blavatsky's "Great Old Ones", Grant's model of extra-terrestrial Gnosis through contact with the being Lam, presented both an alternative model of spiritual enlightenment and an alternative religious cosmology that dealt specifically with 1970's America's fascination with the new frontiers of space, nuclear physics and an the millenarian/apocalyptic discourse of nuclear armageddon during the cold war. Staley in his lecture on Grant's impact on the post-Crowley occult tradition, cites a particular passage by Lovecraft from the short story "The Wall of Sleep" which he felt represented the new cosmology and extra-terrestrial gnosis presented by Grant.

> From my experience, I cannot doubt but that man, when lost to terrestrial consciousness, is indeed sojourning in another and incorporeal life of far different nature from the life we know, and of which only the slightest and most indistinct memories exist after waking. We may guess that in dreams, life, matter and vitality, as the earth knows such things are not necessarily constant; and that time and space do not exist as our waking selves comprehend them. Sometimes I believe that this less material life is our truer life and that our vain presence on this terraqueous globe is itself the secondary or virtual phenomenon.

Lam itself, made famous by a portrait painted by Crowley, was the product of a series of trance oriented magical rituals conducted with a woman by the name of Roddie Minor in New York during 1917 in which they engaged in ritual workings involving magica sexualis, hashish and opium. Together they induced a series of visions of archetypal beings involving a King, a small boy and a Wizard who presented himself as named Amalantrah. The portrait of Lam was believed to have been encountered by

Crowley during this period as a higher being or class of being introduced to Crowley by Amalantrah.

The portrait itself fell into the hands of Grant, after a series of astral visions in which Crowley and Grant were both involved. The only comment attributed by Crowley is an off hand remark to a reporter that the portrait was of his mentor and the phrase he wrote under the painting,

> Lam is the Tibetan word for Way or Path, and LAMA is he who Goeth, the specific title of the Gods of Egypt, the Treader of the Path, in Buddhistic Phraseology. Its numerical value is 71, the number of this book.

Of the portrait itself Grant writes on his website that,

> The Cult of Lam has been founded because of very strong intimations that have been received by Aossic Aiwass, 718 to the effect that the portrait of Lam (the original drawing that was given by 666 to 718 under curious circumstances) is the present focus of an extra-terrestrial – and perhaps trans plutonic – energy which the OTO is required to communicate at this critical period, for we have now entered the 80's mentioned in the "Book of the Law". It is our aim to obtain some insight not only of the nature of Lam, but also into the possibilities of using the Egg [referring to the shape of the Grey's head as an egg symbolically representing the Lapis or rebirth in alchemy] as an astral space capsule for traveling to Lam's domain, or for exploring extra-terrestrial spaces...

Interestingly Grant, who was a fan of the work of H.P. Lovecraft argued that Lam was analogous to Lovecraft's "great old ones". Furthermore, Grant claimed Lovecraft's Cthulu Mythos represented an accurate, albeit metaphorical, representation of the cosmological universe as epitomized by the Lam entity. As he wrote,

> Crowley was aware of the possibility of opening the spatial gateways and admitting an extraterrestrial current in the human life wave. It is an occult tradition – one that Lovecraft gave persistent utterings in his writings – that some transfinite and superhuman power is marshalling its forces with intent to invade and take possession of this planet. This is reminiscent of Lovecraft's dark hints of a secret society on earth already in contact with cosmic beings and, perhaps, preparing the way for their advent. Crowley, however, dispels the aura of evil, with which Lovecraft invests in this fact; he prefers to interpret it thelmically, not as an attack upon human consciousness from within, to embrace other stars and absorb their energies into a system that is thereby enriched and rendered truly cosmic by the process.

In this way contact with alien entities represented a new way of grappling with the unknown and uncertainties of a new world which incorporated a new cosmology in a new decentered chaotic cosmological universe through ritualizing the process of contact and embrace of the unknown. After a series of ritual preparations one was to gaze intently into the eyes of the portrait until entering a trance like self hypnotic state and then open one's eyes to look out through the eyes of Lam onto an Alien landscape. One would then engage in banishing ritual and return to a normal state of consciousness. People would then describe a series of visitations and abductee like experiences of contact with Lam and Lam like entities.

Perhaps the most interesting aspect of Grant's work and influence on the New Age and occult communities is the extent to which there is a process of mutual formation between popular culture, alien mythology and the theosophical and occult tradition. Furthermore, as the Lam statement indicates, it is quite possible to induce contactee-like experiences at will and these experiences created by the self inducement of a hypnotic trancelike state closely resemble that of contactees. Indeed, the parallels of dreamlike shamanistic states linked to deep and invasive penetration of the self, combined with an over thematic structure of psychological development and anxiety seem to coalesce between the two experiences, albeit one that is induced at will through meditational practice, the other, if skeptics are to be believed, is the product of a hypnogogic or hypnopompic state. Both experiences describe a similar sense of dislocation, sexual invasiveness, psychic transformation and deep personal anxiety centered on aspects of the psyche associated with issues of personal anxiety. The same sense of timelessness and liminal isolation from the existent mundane world is also a commonality shared between the two experiences.

One other important issue to contend with in attempting to interpret the social and cultural significance of the shamanic nature of the UFO abductee/contactee experience is the argument posited by occultist Blake that much of the contemporary UFO abductee phenomenon is a product of the popularization of regressive hypnosis during the 1980's. Furthermore, that this methodology, as popularized in Whitley Strieber's book 'Communion' and in part of broader approach to the issue of memory in reconstructing the past that was instigated by the Satanic Ritual Abuse Scare of the 1980s, was combined with the increased popularization of the Alien mythos on literature and cinema. Indeed, this process was, also considerably exacerbated by the complicated matrix of feedback between New Age spiritual practices and beliefs and their representation in popular culture. This is also a point shared and elaborated on by noted skeptic society advocate Michael Shermer who argues that,

> In my opinion, the Alien abduction phenomenon is the product of an unusually altered state of consciousness interpreted in a cultural context replete with films, television programs and science fiction literature about Aliens and UFO's... Driven by a mass media that revels in such tabloid type stories, the Alien abduction is now in a positive feedback loop. The more people who have had these unusual mental experiences see and read about others who have interpreted similar incidents as abduction by aliens, the more likely it is they will interpret their own experiences as Alien abductions.

Once again we are left with the strange ephemeral and trickster nature of the UFO experience and its deep permeation of popular culture. Thus we encounter an ephemeral trickster cosmos in which the tools of science seem much blunter than usual. In this way, as Jung argues, UFOs are a modern myth deeply tied to the cultural context and an expression of spiritual liminal experience outside the perceptual boundaries of time and space. This very liminality of the experience led Edward deBono to comment on the impossibility of a skeptical approach making inroads against such a deeply permeating modern mythology. He writes,

> Myths cannot be destroyed by direct attention since they are the organizing pattern on the memory surface and any attention to a pattern can only re-enforce it. A myth can only be destroyed through inattention which lets it atrophy so that a new organizing pattern can arise. Inattention or neglect usually follow when a myth has outlived its usefulness.

Fundamentally then, the realm of the UFO and the Alien are themes which deeply permeate popular culture and the experiences associated with them and the movements they inspire are intrinsically linked to the underlying cultural and social infrastructure which both underpins and formulates that experience. To this extent, typically, approaches to Alien beliefs and experiences which focus almost entirely on the objective empirical reality and veracity of the experience tend to obfuscate much more than they reveal. Rather, by looking at the experiences psycho-socially and anthropologically one gains a greater grasp of the meaning of the event both psycho-analytically and culturally. In a very pragmatic sense one can examine them as religious phenomenological experiences and evaluate them accordingly in terms of their symbolic representation, the theology they inspire and in their import for the underlying semiotic representations and ideological structures. Without moving prematurely into generic universalist psychological interpretations of the experiences, one can certainly conclude that, like all religious experience, UFO encounters involve complex interactions of cosmos, culture and subjective consciousness. But such an observation also points to the importance for understanding larger issues of religious and supernatural experience in contemporary life and the cultural context forthwith. As Jodi Dean argues,

> However, much as they may embody "high strangeness" or a sense of the uncanny, in the current insider term, or seem merely bizarre and surreal to the skeptic on the street, UFO revelations and abductee experiences and the groups they have inspired are living laboratories of religious and supernatural experience that can teach us much about both religion and the cosmos.

In this context the shamanic and spiritualist quality to abduction experience is of special significance. Rather than being convinced of skeptical arguments and the banality of a rationalized existence, abductees and contactees and their related organizations are fundamentally engaged in a search for meaning. They accumulate evidence, experiences and literature. They form support groups and religious organizations and lobby for belief in both the validity of the experience and its propagation in popular culture. The experience is also fundamentally linked to the process of the globalization of culture, through both the manifestation of shared popular cultural representations and the tendency of analysis to collectively group disparate experiences and cultural forms. Additionally, the experience is interpreted globally in terms of a united earth sharing a collective humanity in relation to the 'Alien' other whether benevolent or malign. It is a world in which we are all connected both culturally and in terms of a shared humanity. It is also, significantly, a view of interconnectedness that does not recognize the mechanisms by which this connectedness is established in terms of economy, technology, transportation or the information revolution. This, in combination with the millenarian, apocalyptic and prophetic nature of the experience, sheds light on many of the concerns and issues facing contemporary society. It also represents a mythological and spiritualist world view which, while recognizing fundamental aspects of a globalized and decentered world order, reconstructs the underlying physical conditions of a globalized technocratic world and its underlying malaise of anomie through the discourse of transcendent religiosity. As Jung comments,

> The basis for this kind of rumor [of UFOs] is an emotional tension having its cause in a situation of collective distress or in vital psychic need. This condition undoubtedly exists today... In the individual too, such phenomena as abnormal convictions, visions, illusions etc, only occur when suffering from a psychic dissociation, that is, when there is a split between the conscious attitude and the unconscious contents opposed to it. Precisely because the conscious mind does not know about them and is therefore confronted with a situation from which there is no way out...

MONSTROUS MONTANA
Michael Newton

Humans love monsters in their proper place--in myths and novels or on movie screens. They offer risk-free thrills and chills, sometimes with comedy relief. We come away from the experience relieved and reassured of our humanity. We have survived.

It is a different story, though, when unknown creatures come to life and pass among us, traipsing through our forests, paddling in our lakes and streams, or peering through our windows after nightfall.

Then the laughter stops.

Montana, as it happens, has no shortage of bizarre creatures. They terrorized Native Americans, greeted the early European pioneers, and they are with us still, waiting to lurch out of the shadows and surprise us when we least expect it.

Big Feet

Don Avery created a sensation in May 2006, while campaigning for a seat on the Flathead County commission. It was not politics, however, but his claim of multiple encounters with the hulking primate known as Bigfoot. Avery told the *Hungry Horse News*, in Columbia Falls, that he'd seen the creature twice in 1996 and found its giant tracks again in 2001. His wife had also seen the beast.

They are not alone.

Montana's oldest and scariest Bigfoot tale was published by future president Theodore Roosevelt in 1892. His book *Wilderness Hunter* includes the story of a trapper named Bauman who ran afoul of Bigfoot in the Bitterroot Mountains, sometime in the early 19th century. The beast trashed Bauman's camp and killed his partner, tearing out the hapless victim's throat. Bauman escaped but left his precious pelts behind while fleeing for his life.

Settlers got even in November 1892, according to the *Anaconda Standard*, which reported the slaying of a hairy "varmint" that was "not unlike a man," seen "walking on its hind legs after the manner of a gorilla." The beast killed and devoured "several large bears and one mountain sheep" before it was shot.

Researcher John Green collected 74 Montana Bigfoot reports between 1900 and 1977. Modern reports to Internet websites boost the state's record to 106 sightings of large apelike creatures unknown to science. Montana's most recent Bigfoot encounters to date were logged from Cascade County in September 2006 and from Big Horn County the following month.

Most witnesses describe Bigfoot as six to nine feet tall, weighing between 300 and 800 pounds, covered in black or dark-brown hair. A pointed bullet-head rests on broad shoulders with no visible neck. Its trademark footprints, humanoid in shape, commonly measure 12 to 19 inches long and 8 to 10 inches wide. Many reports include foul odors and a high-pitched shrieking call.

On June 18, 2003 an unidentified witness allegedly videotaped a Bigfoot near Lake McDonald, in Glacier National Park. While unaware of the beast at the time, the hiker noticed it when he watched the tape and sent a copy to the Bigfoot Field Researchers Organization. BFRO investigators describe the image--never publicly aired--as "a biped" running through woods with "very little detail."

The last report of "hairy men" killing livestock came from ranchers living near the Snowy Mountains, in the early 1900s, but Bigfoot is not to be trusted. Between February 1996 and August 1997, seven Montana witnesses reported aggressive behavior by Bigfeet in three separate incidents, at Gerber, Great Falls, and Belt Creek Canyon.

Trappers beware.

Lost and Found

In 1886 a mysterious predator known as *shunka warak'in* ("carrying off dogs" in the Ioway dialect) ravaged livestock herds in the Madison Valley. Rancher Israel Hutchins shot it and gave the carcass to taxidermist Joseph Sherwood, who displayed it for years in his store at Henry's Lake, Idaho.

The creature vanished when Sherwood's collection passed to the Idaho Museum of Natural History, but descriptions and a single photograph remained. Those who saw the beast said it was four feet long (excluding tail) and stood 28 inches tall at the shoulder. Its form was vaguely dog-like. Some believed it was a mutant wolf; others suspected an "escaped hyena."

While the photograph of Sherwood's specimen survived, labeled "Guyasticutus," the subject's strange appearance frustrated all efforts to identify it. With the carcass lost, all theorists agreed, solution of the

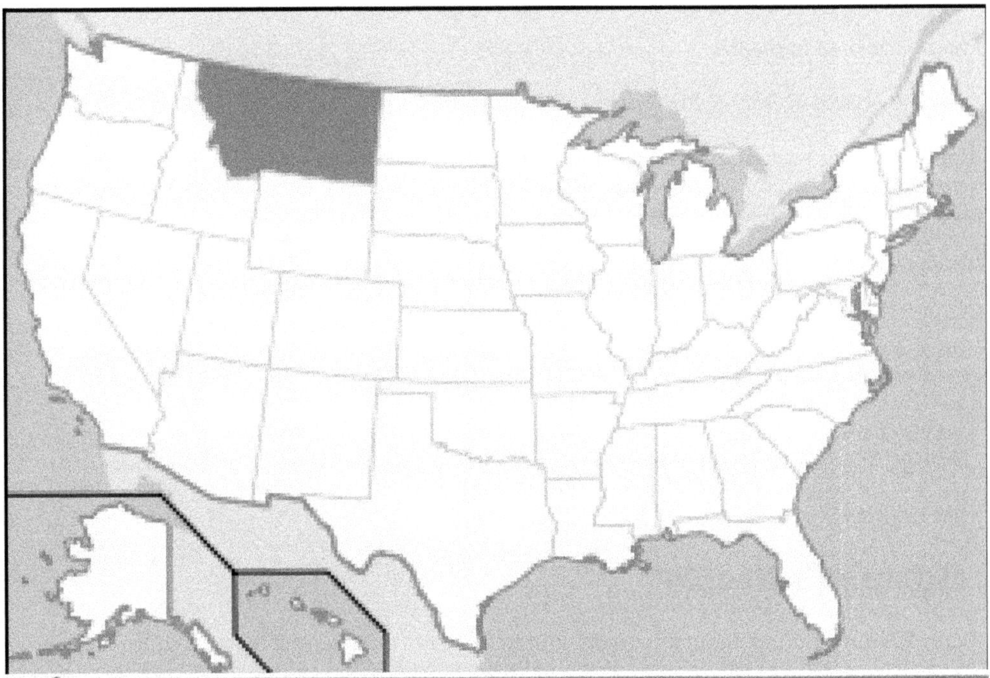

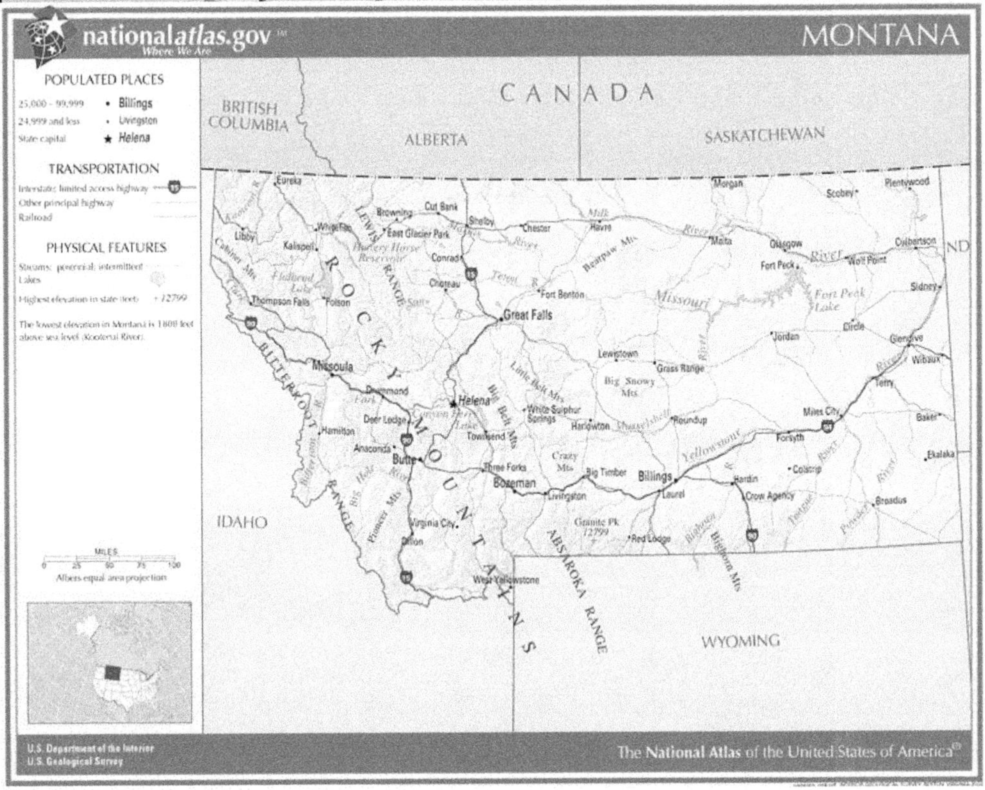

nationalatlas.gov
Where We Are

MONTANA

POPULATED PLACES
- 25,000 – 99,999 • Billings
- 24,999 and less • Livingston
- State capital ★ Helena

TRANSPORTATION
- Interstate limited access highway
- Other principal highway
- Railroad

PHYSICAL FEATURES
- Streams: perennial; intermittent
- Lakes
- Highest elevation in state (feet) ▲ 12799

The lowest elevation in Montana is 1800 feet above sea level (Kootenai River).

MILES
0 25 50 75 100
Albers equal area projection

U.S. Department of the Interior
U.S. Geological Survey

The **National Atlas** of the United States of America

mystery remained impossible.

Until it resurfaced in October 2007.

At last, 121 years after Israel Hutchins fired the shot heard 'round Madison Valley, Idaho museum curators found the stuffed carcass gathering dust in storage. They loaned it to Jack Kirby, Hutchins's grandson, for display at the Madison Valley History Museum, beginning in May 2008.

But what _is_ it?

Simple viewing of the beast, with its black-and-tan hide and suggestion of stripes on the sides, fails to answer the question. Researchers hope that a DNA test may resolve the riddle, but they've hit an unexpected roadblock with Jack Kirby.

It seems that Kirby cherishes his clan's enduring mystery, asking Walt Williams of the *Bozeman Daily Chronicle*, "Do we really want to know?"

Some of us do. We do, indeed.

"Montana's Nessie"

Nearly 900 lakes worldwide produce reports of large unknown creatures, but Montana's Flathead Lake is the only one said to host *two* different species of monsters. With a surface area of 191.5 square miles and a maximum depth of 370.7 feet, Flathead is the largest freshwater lake in the western coterminous United States - and its depths spawned strange legends before the first white man arrived at its shores.

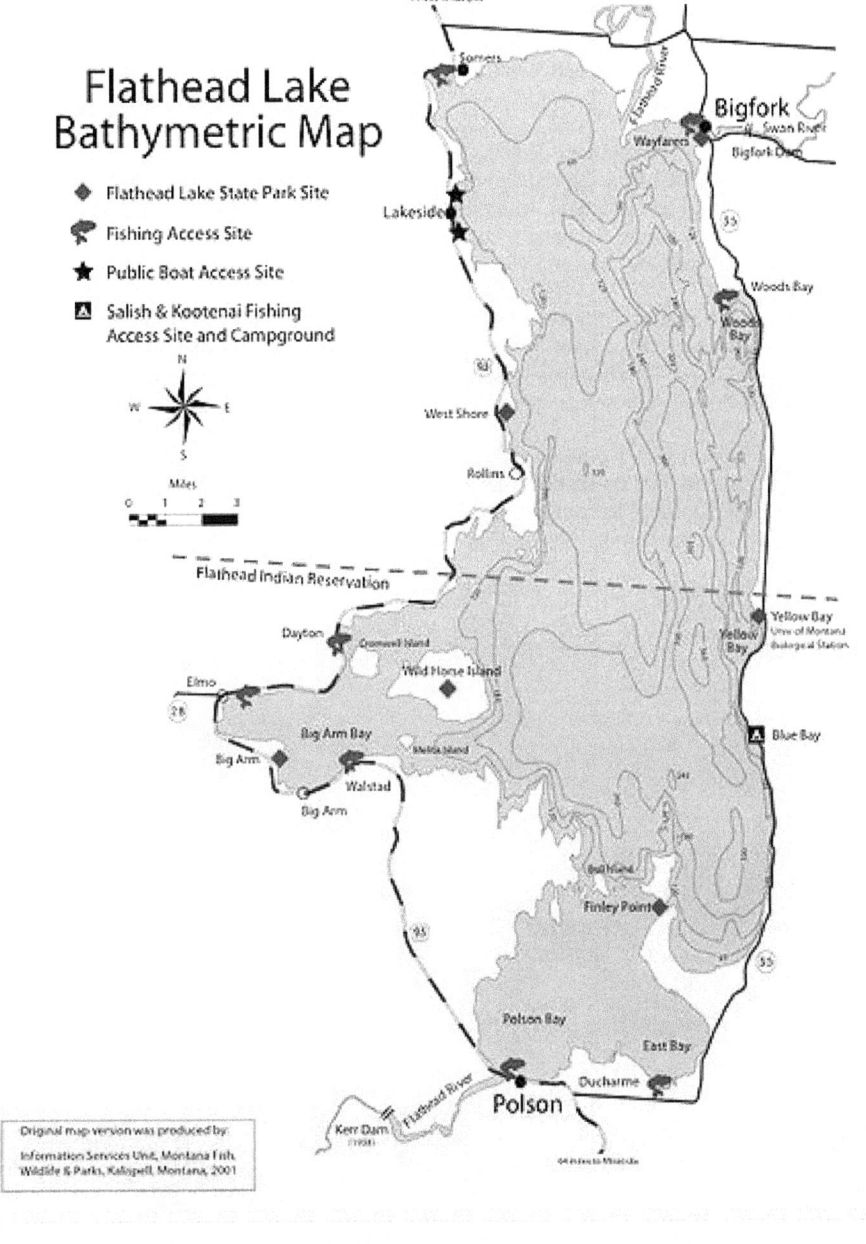

Kalispell Indian legends describe an aquatic monster that raided tribal camps around Flathead Lake in prehistoric times. The first known Caucasian witness was Capt. James Kerr of the steamer *U.S. Grant*, who met a whale-sized creature on the lake in 1889. One of Kerr's passengers fired a rifle at the beast, which caused it to submerge. Thirty years later, another steamboat dodged a "living log" on Flathead Lake, and sightings proliferated thereafter - at least 90 by 1998, according to Paul Fugleberg, editor of the *Flathead Courier*.

Spokesmen for Montana's Department of Fish, Wildlife, and Parks admit logging 78 monster reports from Flathead Lake, but their files suggest two different species of creatures. One-third describe a giant fish, ranging from 6 to 20 feet in length, while the remainder sketch a larger eel- or snakelike beast resembling monsters reported from Scotland's Loch Ness and elsewhere. Predictably, the second creature has been nicknamed "Flattie" or "Montana Nessie."

What dwells within the depths of Flathead Lake? Some researchers believe the giant fish are sturgeon. The largest specimen on record, caught by Russian fishermen in 1827, measured 29 feet 6 inches and weighed 3,359 pounds. A 19th-century biologist, Karl Ernst von Baer, claimed proof from "direct observations in Russia" that individual sturgeon may live up to 300 years.

On May 28, 1955 fisherman C. Leslie Griffith allegedly hooked a 181-pound sturgeon, 7 feet 6 inches long, in Flathead Lake. He hired a flatbed truck to haul his catch away, and while the mounted fish remains on display at the Polson-Flathead Historical Museum, some ichthyologists dispute its provenance, insisting that Flathead Lake harbors no sturgeon. Since no other sturgeon has ever been caught at the

lake, the argument endures.

Whether a colony of giant fish exist in Flathead Lake or not, they fail to account for the 40-to-60-foot serpentine creature seen by George Cote and his son at Yellow Bay on May 25, 1985. The beast had a "head and body resembling a snake, with a tail of an eel." Cote told state wildlife officers, "At one point its head came up high out of the water and it appeared to be looking at us. I counted six to eight coils of its body on the surface but couldn't see its tail because it was under water."

Two years later, in July 1987, the Cotes saw the same beast or its twin near Lakeside. George later wrote, "This one was quivering as it swam, forward motion....[W]e saw the entire head, body and tail."

Flattie logged its last public appearance on August 18, 1998, but its legend endures--with a twist. While monsters worldwide are depicted in art, as souvenir toys, and in TV commercials, Montana's Nessie has spawned its own line of Flathead Lake Monster Gourmet Soda, available in four fruit flavors from the North American Beverage Company in Ocean City, New Jersey.

They quench the thirst, but do not solve the mystery.

SOURCES

Bigfoot Field Researchers Organization, http://www.bfro.net/GDB/state_listing.asp?state=mt.
Chad Arment, *Historical Bigfoot* (2006).
Janet and Colin Bord, *Bigfoot Casebook Updated* (2006).
Loren Coleman and Jerome Clark, *Cryptozoology A to Z* (1999).
Peter Costello, *In Search of Lake Monsters* (1974).
George Eberhart, *Mysterious Creatures* (2002).
Betty Garner, *Monster! Monster!* (1995).
John Green, *Sasquatch: The Apes Among Us* (1978).
John Kirk, *In the Domain of the Lake Monsters* (1998).
Michelle McGovern, "Montana Nessie," http://www.montana.com /territories/aboutmt/index.php3?uid=94&morestories=1.
Michael Newton, *Encyclopedia of Cryptozoology* (2005).
"Sturgeon," Wikipedia, http://en.wikipedia.org/wiki/Sturgeon.
Walt Williams, "Mystery monster returns home after 121 years." *Bozeman Daily Chronicle* (Nov. 15, 2007).
"World Records: Sturgeon," http://www.fishing-worldrecords.com/record_length_01.htm and http://www.fishing-worldrecords.com/record_weights_01.htm.

BELLISSIMO BEASTS: CRYPTOZOOLOGY ITALIA
Neil Arnold

'To him that watches, everything is revealed' – Italian proverb.

When Italy lifted the World Cup aloft in 2006, my heart beat fast, my eyes filled with joyful tears and my love of Italian football was justified after years of it being considered a negative, regressive and cynical brand, despite it having produced some of football's greatest players including Roberto Baggio, Francesco Totti, Paulo Maldini, Franco Baresi, Gianni Rivera and Gigi Meroni. Of course, many of these names remain out of the knowledge of your average football fan who remains drowned in today's commercialism and high drama of players called 'great' but which are nothing more than overblown primadonna's.

Why am I telling you this ? Well, like its wonderfully tactical football, its beer of the highest quality, being Nastro Azzurro, Morretti and Messina, and its perfect production line of scooters, Italy has always been overlooked, despite our fascination with gangster films and the wonderful food. The cryptozoological aspect is another of the countries fine but understated qualities.

In the June of 2008 it was reported that the myth of the Unicorn had becom a reality when a deer, harbouring a single horn on its forehead was spotted at a nature reserve in Tuscany. The animal, a one-year old roe deer born in captivity attracted crowds galore to its genetic flaw which Gilberto Tozzi, the director of the Centre of Natural Sciences in Prato, commented as being possibly reason as to why we believe in the fantasy of the mythical creature known as the Unicorn. Although single-horned deer are rare, the positioning of such a horn is.

Of course, this incredible discovery hit the headlines in the same way the national team's World Cup victory did. Fleetingly. However, Italy, in Southern Europe, and bordered by France, Austria, Slovenia, Switzerland and the independent states of Vatican City and San Marino, is known for its rugged terrain, rustic beauty and chiselled population of over fifty-million.

Within its boot-shaped peninsula, flanked by the Adriatic and Tyrrhenian Seas and hazed by a diverse climate, the country is allegedly inhabited by many a beast which fits into the term known as 'cryptid'. Despite its large cities, mainly the capital Rome, Milan and Naples, attracting millions of tourists each year wildlife is abundant and maintains its mystery, but mainly in its waterways.

WATER MONSTERS

The northern lakes are situated at the foot of the Alps. Lake Como is said to be the deepest, and Lake Garda the largest, although the Apline scenery enshrouds other lakes known for their history and myth, these being Iseo, Maggiore and Orta, whilst elsewhere in the country are Lake Viverone, Lugano, Endine, Idro, Santa Croce, Auronzo, Revine and Lago, Misurina, Alleghe, Pieve di Cadore, Caldonazzo, Levico, Tovel, Lavarone, Toblino, Ledro, Molveno, Trasimeno, the lakes of Varese's Territory, lakes of Avigliana, and in Latium, the north of Rome, situated Lake Albano, Bolsena, Bracciano, Nemi and Vico respectively. Many of these murky mirrors are known for their monster legends. Lake Maggiore, Italy's second largest lake, is said to be inhabited by a serpentine form, it is Italy's most famous lake monster. Despite the variety of water sports which take place on the lake, the local beast has made a small name for itself although remains reclusive in the three-hundred metre depth. Sightings which have taken place describe a horse-headed creature although reports of the length vary. A creature with a head like a horse was seen in the River Ticino in 1934, this northern Italian river actually links into Maggiore and *could* also explain some reports from southern Switzerland. Travel author Stendhal allegedly

wrote of the Maggiore monster at the beginning of the nineteenth century although no trace has been found of the notes, and it is also rumoured that only a handful of locals around the area know of the monster.

In 1946 a large sturgeon was caught in Lake Como but the following year police were said to have intercepted also, a mystery submarine used by smugglers so maybe either, or both of these possibilities could well have explained the resident leviathan. A rather vague legend surrounds Lago Specchio di Venere from Isola di Pantelleria, Sicily, a creature said to have been heard but not seen in 1982, whilst in the 1930s legend was rife in Perugia (north of Rome) that a marsh monster was prowling the vicinity close to the sea by the Tiber River. Janet and Colin Bord briefly mentioned the strange beast in their Unexplained Mysteries of the 20th Century book. The couple also mention a ten-feet long snake with legs said to inhabit Po di Goro from the Emilia Romagna region. The creature was allegedly sighted on 27th June 1975 but experts and some locals believed the 'monster' was nothing more than an escaped crocodile. A similar creature, but without legs was reported in The Times of London of 1933 from Sicily with one specimen allegedly being captured and killed in a swamp near Siragusa. Villagers believed that the creature was a bad omen so by the time journalists were on the scene the carcass of the mystery animal had been burned. Folklore states that such a legendary snake be known as the Colovia.

A creature resembling an eel was sighted in the summer of 1958 by tourists spear fishing in the Copanello region. The sighting was only a brief episode as the witnesses were so terrified by the monster that they fled to their boat. In his book *Dragons: More Than A Myth ?* author Richard Freeman wrote of a dragon-like monster stating, "From the 1930s to as recently as the 1980s, basilisk-like creatures have been reported in Italy. These include an eight-foot green and yellow 'dragon' seen near Monterose – north of Rome in 1935. One local man claimed to have seen it every ten to fifteen years since he had been a boy. Near to the town of Flori, a farmer reported being chased by a fifteen-foot lizard with searing hot breath. The beast chased him for two-hundred yards, as he ran for his life – its searing exhalation at his heels."

Freeman also goes on to write of the already mentioned beast from Goro in 1975 which appeared as a ten-foot long snake. A farmer spotted the creature and reported its presence to the police but when he returned with the authorities the form was gone. In relation to earlier reports and from the Mediterranean sea, during the 1740s a group of fishermen claimed to have spotted a giant eel-like creature at Sarica, Sicily, and in 1741 near Licata, Sicily, a huge monstrous specimen was found beached. The 'fish' was said to measure over forty-feet and over twenty-three feet in circumference and had two large dorsal fins. The stranded creature had a blowhole, a mouth full of razor sharp teeth and defied scientists. An author named Samuel Rafinesque in 1814 gave the monster a Latin name, *Oxypterus mongitori* yet no similar fish has revealed itself, and it wasn't until 1877 that another creature, this time displaying several fins, was seen off Sicily, this time at Cape Vito on June 2nd. Captain L. Pearson and officers of Royal Yacht Osbourne observed the leviathan. In his fascinating book *The Leviathans*, Tim Dinsdale covers the sighting, stating that Commander Pearson, in forwarding his report to the Lords Commissioners of the Admiralty, stated 'I myself saw the fish through a telescope, but at too great a distance, about 400 yards, to be able to give a detailed description; but I distinctly saw a seal-shaped head, of immense size, large flappers, and part of a huge body.'

Two other officers who also observed wrote of their accounts also, a Lieutenant Haynes wrote, '…my attention was first called by seeing a ridge of fins above the surface of the water, extending about 30 feet, and varying about 5 to 6 feet in height…I distinctly saw a head, two flappers and about 30 feet of an animal's shoulder. The head, as nearly as I could judge, was about 6 feet thick, the neck narrower, about 4 to 5 feet, the shoulder about 15 feet across and the flappers each about 15 feet in length. The movements of the flappers were those of a turtle, and the animal resembled a huge seal…', whilst Lieutenant Douglas M. Forsyth concluded, 'The animal was slowly swimming in a westerly direction, propelling itself by means of two large flappers or fins, somewhat in the manner of a seal. I also saw a portion of the body of the animal, and that part was certainly not under 45 or 50 feet in length.'

One of the earliest ever reports of a lake monster in Italy emerges from the 6th century from the Tiber River. A beast resembling a huge piece of wood was observed near Rome after a flood. Maybe, as it was after a flood, it was simply just a piece of debris! Another early report, this time from 1651 claims that near the city of Rome a dragon once inhabited a large cave. In the San Marco Church, Milan, there is said to exist a painting of Knight Visconti slaying a dragon-like creature, an alleged true account from the Middle Ages. Strangely whilst it has been claimed that the creature on the end of the knight's sword is a dragon, it actually resembles a dinosaur, such as a small diplodocus. Unfortunately, the painting was damaged during the Second World War and so the creature in the painting is bereft of a head. However, one of the strangest pieces of evidence to turn up in Italy regarding unknown serpents was stumbled upon on February 10th 1968 by a fisherman named Vincenzo Croce, who, whilst taking a stroll along a beach near Campobello, in Sicily, found a large bone sticking out of the sand. This was no discovery of a prehistoric creature that once roamed the planet but in fact the remains of a beast that hadn't been dead that long as pieces of flesh still remained attached to the framework of the rest of the bones which Vincenzo unearthed. The form was more than twenty-five feet in length, had a duck-shaped head, and before it was transported to a museum, many tourists flooded the area to stare at the inspiring goliath from the sea. At the time, scientists refused to say that the bones were from a monster but also couldn't connect the remains to any known sea creature. Meanwhile, one of the silliest hoaxes in Italian history revolved around an alleged serpent. An undated commercial holiday expedition into the forests of Palermo, resulted in several holiday makers fleeing in terror from a creature that was said to have resembled a Plesiosaur. A Ben Grace, who was reported to have wet himself after seeing the beast told journalists, "I remember the appalling skin of the creature, like paper mache over a chicken wire frame." It seems that the local monster was created for a bit of fun and has done the rounds on the internet.

Mythology speaks of a fearsome beast said to inhabit the region between two rocks named Scylla and Charybdis, at the Messina strait. The creature was said to have twelve legs, six heads, each owning jaws consisting of razor sharp teeth and the monster was said to bark like a dog. A creature known as Silvester's Dragon was said to have been slain in Roman mythology. The fiery critter had a lair at the foot of the Palatine hill. The creature had killed many innocent people and so Pope Silvester I decided to

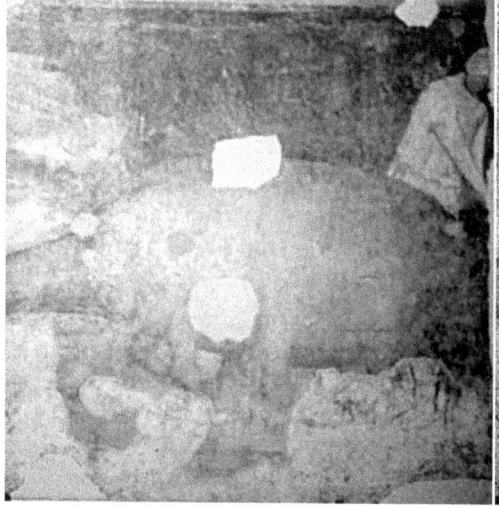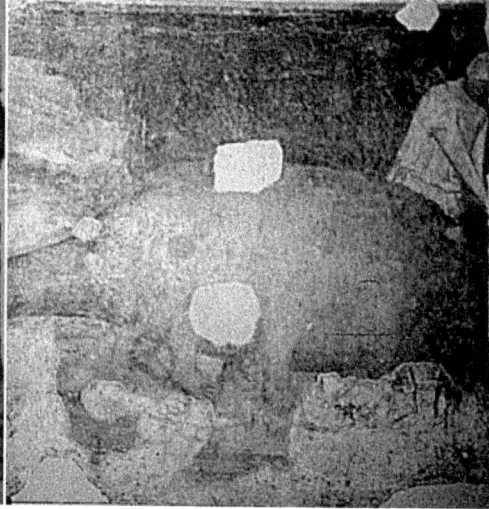

put an end to the bloodshed and tamed the beast with crucifix alone. Legend states that by thread he lead the leathery-skinned dragon across the land, handed it over to some unknown clan, who disposed of the menace. At a medieval church in Bergamo, it is said that an unusual bone, known as 'Dragons Rib' hangs from the ceiling. The bone is said to be part of some mystery beasts rib-cage, and measures 1.70 metres in length. In 1953 a monstrous set of footprints were observed by a sea fisherman in the vicinity

of Marina di Pisa as he strolled the beach looking for a location to fish for molluscs. Bizarrely, the huge impressions he discovered were coming from the sea. The prints eventually went out of sight once they reached a pine forest.

Finally, the quirkiest 'dragon' story from Italian mythology originates from the 1642 work *The Natural History Of Serpents & Dragons* in which scientist and author Ulysses Aldrovandus speaks of a weird encounter, in which he states, 'The dragon was first seen on May 13, 1572, hissing like a snake. He had been hiding on the small estate of Master Petronius near Dosius in a place called Malonolta. At 5 PM, he was caught on a public highway by a herdsman named Baptista of Camaldulus, near the hedge of a private farm, a mile from the remote city outskirts of Bologna. Baptista was following his ox cart home when he noticed the oxen suddenly come to a stop. He kicked them and shouted at them, but they refused to move and went down on their knees rather than move forward. At this point, the herdsman noticed a hissing sound and was startled to see this strange little dragon ahead of him. Trembling he struck it on the head with his rod and killed it.'

The carcass was examined and mounted for a museum. The creature had a long neck, a rotund body and very long tail, suggesting some kind of Monitor lizard perhaps? Seemingly not. Aldrovandus noted that the creature appeared harmless and was a juvenile but was unlike anything he had ever seen because it only had two front legs and manoeuvred by a slithering effect from its snake-like rear end. The slender neck of the beast had white markings which circled and the animal when reported alive was said to often raise its head and hiss like a snake.

Could such an unknown creature have once inhabited much of Europe and explained a lot of the dragon legends? Also, the tatzelwurm, a strange creature from the Swiss mountains was also said to prowl into Italy's rugged landscape. Such a monster was sighted in 1894 near Merano in the north and described as measuring seventy-centimetres in length, having no legs but a slithering body being the thickness of a mans arm, a blunt head, and overall grey in colour. The creature was killed by a hammer. Forelegs were described on a Tatzelwurm from Austria which was encountered in 1916. The critter was said to have frog-like eyes, but all other details were similar to the Italian account. In general, the legend of this mystery prowler suggests that the Tatzelwurm resembles a cross between a long-bodied cat, salamander, snake and lizard. The Tatzelwurm is best observed by keeping still and remaining out of view.

Sadly, much evidence of lake monsters splashing around in Italy's lakes is scant. Our only other hope is that some sea-dwelling monster washes up on a beach as in the case of this Leatherback Sea Turtle which became stranded in Genoa several years ago, or the extremely large wels catfish which lurk in the murky rivers become even larger. One of the largest specimens of this wide-mouthed fish ever caught was from the Po Delta at 144 kg, and measuring 2.78 metres.

MYSTERY FELIDS

Put simply, I think it would actually be more of a mystery to find a place in the world that isn't inhabited by mystery cats. Although Italy sits in the lower leagues in regards to 'big cat' sightings, over the years a few slinking felids have been reported, mainly those being black in colour which are also being experienced in numbers from the British Isles to the United States of America and Australia, suggesting a mixture of massive feral cats and melanistic leopards. However, some sort of panic hit Turin around 2005 and into 2006 with much of the blame pointed at a Russian Zoo situated in the area although spokesmen for the exhibition stated categorically that they didn't have any cats, but only housed dogs, camels and horses. However, a traffic officer claimed to have seen a black 'panther' in the Piedmont metropolis but after several searches the local authorities found no trace. The panic echoed similar flaps in the Umbria region from 1993 and 1995, as well as the Como panic of 1999 and Sardinia in 2001. In 2000 Italian ufologists (of all people!!) began investigating black 'panther' reports after livestock was found mutilated and killed in the province of Tarquinia, sixty-miles north of the capital. Website UFOInfo wrote, '"In Tarquinia, numerous reports of a ferocious animal, said by some to be a panther, have circulated." In late February, "…a cow was found dead in a field 1.5 kilometres from a stock breeding farm located just south of Aurelia Bis. The animal had been in the field for two or three days. The flesh had been stripped from its bones, and all that remained was the spine. People in the area continually talk about the problem."

"The last sighting was on the border with Tuscania (Italy's Tuscany province--) Umberto Angelini was horseback riding in a farm area near Pisciarello when he spotted the strange creature a considerable distance away."

According to Commandante Amorevole of the local Carabineri (Italy's national police-), "'Angelini was a good distance away, not in close proximity to the animal. He was about 400 meters away from the field, and he saw a great black animal--a beast able to rip the flesh right off the carcass of any animal weighing 60 kilograms and carrying it away..''

The Carabineri "…excluded the possibility of the animal being an escaped lion or tiger. They also excluded as the culprit a selection of other known predators." (Grazie a Alfredo Lissoni di Centro Ufologico Nazionale d'Italia per questo rapporto.)

In 1992 there were reports of a 'phantom panther' around Avellino, near Napoli. Then, in the haze of June a similar cat was seen at Colfiorito, followed a month later by a sighting in the Brindisi district and a month after that in Treviso. Strangely, from the same source, a Paolo Fiorini commented on other animals being released into the wilds of Italy, stating briefly, 'Since several years, odd rumours have began to spread throughout Italy: some species of animals, bred in captivity or purloined, would be carried by night by means of trucks in order to be unlawfully let in several areas. The operators remain unidentified: some say hunters, some other say ecologists, or landowners, or nothing less than the legal authorities charged with the safeguard of natural patrimony. We give three examples regarding alleged "unlawful" removal of pigeons, deer and wild boars in some areas of Piemonte.'

In 1993 Paolo Toselli, under the headline THE PANTHERS' INVASION commented, 'During the whole 1993, there have been dozens of sightings of "mystery cats" all over Italy, to a great extent in central and north-western countries. It seems that panthers, leopards, pumas and lions are undisturbed running about our woods as well as our downtowns. We've got torn to pieces animals, quick sightings, some prints, a couple of photographs and continual beatings. Evident captures are virtually non-existent. To tell the truth a lioness and a puma would be captured by a screen actor in Lazio, but it was almost surely a hoax for advertising reasons. A small part of sightings of such big cats are likely due to animals left behind in the country by their owners. Anyway if we look at it as a whole, the phenomenon is an interesting example of contemporary legend.'

Ten years previous, and mentioned in Karl Shuker's *Mystery Cats Of The World*, two puma were allegedly released in Bari

In 2006 source Ansa reported PANTHER PANIC STRIKES AGAIN - Big game hunt launched in northern countryside

Piacenza - Italy has been hit by one of its periodic bouts of panther panic with a full-scale 'big game hunt' in the countryside around this northern city. Authorities said they had sent out police and forest guards after several sightings of "a big cat, probably a panther" in the local hills. They said the animal may have escaped from a circus "or the villa of some eccentric citizen" .

The latest feline flap followed a similar scare on the outskirts of Rome in February, when panic set in after six goats were found dead . Locals calmed down when experts said the tooth marks on the goats were those of a big dog. The Piacenza hunt is Italy's seventh panther alert in just over a decade . In December, Turin was shaken by panther fears after a series of sightings sparked rumours that a visiting Moscow zoo had lost one of its biggest draws .

The zoo squashed the reports. The December reports revived memories of a wild cat chase near Turin six and a half years ago .That 1999 report spurred copycat incidents as feline fever spread to the provincial border with Como .

Most of the other reports of predators on the loose have come amid midsummer news droughts known to journalistic insiders as "the silly season." Panthers were reported to have been sighted in the harsh landscape of central Sardinia in 2001 and the rolling countryside of central Umbria in 1993 and 1995. After the second Umbria sighting locals said they had also seen a lion roaming in the vicinity.'

Around 2000 I spoke with a then work colleague who mentioned that whilst visiting his wife's family in Italy (location unknown), they were driving one evening when a large, fawn-coloured cat emerged ahead on the road and slinked off into the undergrowth. The witness, a sincere man named Bernard said he may have seen a lynx which whilst on the subject of, I'd like to reproduce a fascinating article in regards to, from 1991 written by Franco Tassi, Professor of the Appenine Ecological Study Centre, Abruzzo national Park, in Italy. To quote:

CAT NEWS: Issue 15, Autumn 1991

The lynx, that mysterious and fascinating animal of the European forests, is in the process of returning to Italy. After almost disappearing from the entire European continent, and after several countries adopted measures for its reintroduction, this extraordinary cat, known in Italy as the "cervine wolf", is now gradually reappearing in the Italian Alps. More substantial human intervention, however, will be needed for its reintroduction into the Apennines.

Information about this predator and its history has been both scarce and contradictory. Scholars such as Ghigi and Toschi considered the lynx totally extinct in Italy as far back as the beginning of this century. Moreover, they excluded the possibility that it lived in the Apennine region, at least within the recent past. But there is little evidence to support these two statements.

Not only are the lynx's past history and present distribution in doubt, but the definition of its systematic position is also unclear. Attention is focused at present, however, on projects for the lynx's reintroduction, which are finally starting to take shape in Italy.

For more than half a century, zoology has officially ceased to consider the lynx among Italian fauna. But there is some doubt as to whether this means the total absence of the cat from our country. The subject has been riddled with misconceptions and become the object of irrational controversies.

Louis Lavauden, French author of the most complete monograph on the European lynx, wrote as far back as 1930: "This animal has been ignored by the populations who lived in contact with it and has left no trace of itself in the folklore of our mountain regions. Its total disappearance from our territory has been asserted with extraordinary perseverance, and the most illustrious zoologists have written genuine absurdities about the lynx".

The Eurasian lynx *(Lynx lynx)* was widespread in Europe, but it is now confined to northern and eastern regions, except for specimens reintroduced into Switzerland, France and Austria. A smaller species, the pardel lynx *(Lynx pardina)*, adapted to the Mediterranean maquis region, exists in the Iberian peninsula. The pardel lynx used to be considered a simple subspecies, but today it is usually regarded as a full species. It was reported in the past from the Italian and Balkan peninsulas, but modern scholars think that this was a southern variety of the European lynx, with more visible spotting.

The Apennine lynx represents the remnants of the European lynx's great southern expansion in the post-glacial period. Therefore it follows that this cat penetrated along the Apennine spine during the Quaternary, together with all the flora (particularly the forests) and the fauna (including its usual prey from the northeastern areas) with which it is ecologically closely linked. Pleistocene evidence from central Italy clearly indicates the presence of the European lynx, which would thus represent the original central Apennine variety, preserved until today.

Only a few years ago, in the summer of 1978, the academic world raised objections to the possible reintroduction of the lynx in Italy, mainly stemming from the belief that there was no protected area in the Alps or the Apennines large enough for lynx. With the same reasoning one could have excluded the possibility of survival of the brown bear *(Ursus arctos)* and the Apennine wolf *(Canis lupus)*, two mammals that not only exist but are actually showing a slight increase in numbers. In any case, the lynx has successfully returned to the Alps, and there are some people prepared to maintain that it may never even have left them. Nowadays the reappearance of the lynx in the eastern Alps is considered a fact, though there had been several sightings in Alto Adige before 1970 and a hunting magazine mentioned its existence in Carnia in 1968. Probably coming from Yugoslavia, the lynx is occasionally seen in Friuli and Venezia Giulia, and is definitely located in the Lagorai-Vanoi region of the southern Dolomites. There were documented cases of illegal shooting of lynx in 1982, 1984 and 1989.

As regards the western Alps, new and interesting information has been revealed by thorough studies carried out by Toni Mingozzi. Even though the lynx has generally been considered extinct there since 1930, reliable evidence has indicated two more recent captures, one in 1937 and one as late as 1947. A possible sighting in Valdieri in 1934 could be added. It is mentioned in the Royal Hunt Register. Mingozzi also revealed a series of sightings in Monviso between 1981 and 1986, which were backed up by observations made by Hainard on the French side of the same mountain range in 1977. Undoubtedly, since 1980, lynx sightings have increased in Italy as well, thus sustaining the hypothesis that the lynx is slowly penetrating from Switzerland. The sightings known to Mingozzi, apart from Monviso, refer to the Val d'Aosta, Val d'Ossola and Valtellina. To these we could also add a sighting made by Boris Zobel near Oulx, in Val di Susa, in the winter of 1981.

A project to reintroduce the lynx in the Gran Paradiso National Park in 1975 fell through when two male specimens were released but their female counterparts failed to arrive. As far as the Apennines are concerned, the lynx's situation is even more uncertain, controversial and delicate. There are no biological, ecological or biogeographical barriers to its former presence in these mountains, primarily in the central-southern area. This is dramatically confirmed by a large number of historical, ichnographic, bibliographic and topological elements, some of which I have already discussed in a work published in 1971 by the *Societa Italiana di Biogeografia*. The lynx's presence in the Apennines has been accepted by such scholars and zoologists as Pietro Doderlein, Emilio Cornalia, Achille Costa, Oronzio Gabriele Costa, Leonardo Dorotea, Erminio Sipari and Alberto Simonetta. However it has not been proven by tangible evidence, such as skins, skulls or skeletons. This may be due to limited research or to the poor hunting and naturalist traditions of our South. Nonetheless, since 1971, I have been able to find additional evidence to support the lynx's presence in the Apennines. Some of this is of great interest and mainly unpublished, and thus earmarked for future scientific publications. I will postpone mention of the evidence on recent sightings in Marche, Abruzzo, Molise, Basilicata and Calabria between 1968 and 1990, which need different tests and research. I will limit myself to only three of the more significant pieces of information out of those which emerged after my work in 1971:

In John Murray's *A Handbook for Travellers in southern Italy* (London, 1855), the lynx, called "gattopardo" (spotted cat) by Abruzzo peasants, is mentioned as living in great numbers in the Marsican mountains. In *Viaggio nel Regno di Napoli* (Travelling in the Kingdom of Naples) by the German writer Carlo Ulisse De Salis Marschlins (1789), an entire chapter of extraordinary interest describes the lynx of the Apennines as an animal well known at the time and easily bred by the Tomassetti barons of Pescina. In the manuscript of the *Fauna del Regno di Napoli* (Fauna of the Kingdom of Naples) by Oronzio Gabriele Costa in 1839, revised by Dr Eugenio Bettoni and now in the Natural History Museum in Milan, the following is written about the lynx: "After publication of the catalogue, Costa received three lynx skins. Subsequently, in a report to the Minister of the Interior, the Superintendant of Chieti declared that a female of the species had been killed in Borello, province of Lanciano."

This event is also reported by C. Lopez, who mentions it as having occurred "a little after 1845".

Finally, the presence of the lynx in the Apennines represents a typical "cryptozoological" case, which has many analogies with other similar ones (for example, the elusive mountain lion of the eastern United States (Felis concolor). This does not mean it should be set aside because of the lack of tangible evidence. Several European countries have for many years been working to bring this splendid cat back to the habitat from which it had disappeared for various reasons. During the past few months, specialists Ulrich Wotschikowsky and Gotz Kerger have reviewed various programmes, implemented between 1970 and 1990, for the reintroduction of the lynx in Europe (not taking into consideration illegal attempts, of which at least four are known, two of which failed). Nine projects have been analysed, three of which met with a high degree of success: two in Switzerland (Obwalden and Jura), and one in Yugoslavia (Slovenia). For two more projects, one in France (Alsace) and the other in Czechoslovakia (Sumava), it is still too early to attempt an evaluation of the outcome. On the other hand, the remaining four, in Austria (Styria), Germany (Bavaria), Switzerland (Waadt) and Italy (Gran Paradiso) have not had much luck. Poor preparation of some of the vital components such as research, management, legislation, risk of damage to the zootechnical patrimony and public relations are among the various reasons for failure of these projects. But sometimes failure is also the result of the limited number of specimens released, or of the choice of habitats (often protected, at least in theory, but not large enough or too snowy). Experts have identified two fundamental requirements for implementing any future reintroduction programme:

1) a natural habitat of proper dimensions, capable of sustaining a self-sufficient population.
2) the elimination of the causes that initially led to the extinction, most importantly deforestation and poaching.

They also recognise that, regardless of expertise and preparation, the success or failure of a repopulation programme may also depend on luck and unforeseeable circumstances. Notwithstanding the many difficulties faced, the conservationist spirit and the dedication of European colleagues have ensured that the future of the lynx appears brighter than it did only 20 years ago. The success in the Alps has been greater than the most optimistic expectations (we have seen how the lynx has undertaken the repopulation of even the most southern part of this great mountain range). In Germany, two additional reintroduction programmes, one in the Black Forest and the other in the Palatine Forest, have been proposed. A proposal to reintroduce the lynx in the Abruzzo National Park has been considered for some time now. The park is the heart of a vast system of protected areas, linked to Maiella, Gran Sasso, Monti della Laga, Sibillini, Monti Ernici and Simbruini, in the green core of the Central Apennines. Scientific, socio-political and organizational confirmation are now awaited to facilitate its implementation. The idea has provoked enthusiasm and hope and is generally seen as a good way to increase the value of the environment, thus completing the reintegration of those ecosystems where other predators are already present, among them the Marsican bear and the Apennine wolf.

Objections to this proposal can be reduced to four salient points:

- the absence of evidence or reliable clues to the former presence of the lynx in peninsular Italy;
- the lack of available space vital for cat population expansion;
- the risk of the lynx preying on rare and localized fauna;
- the possible hostility of local people.

In reply, it should be pointed out that:

- there are consistent traces of the historic presence of lynx in peninsular Italy, no less than those relating to other species in the same time and place. It would be unethical to

refuse this evidence a priori, since in similar cases it has been willingly accepted;
- the overall space available in the Apennine system of protected areas is sufficient, as shown by the presence of large predators in those areas and by the spontaneous return of the lynx to areas certainly less protected and far smaller;
- the possibility of the lynx preying on the Abruzzo chamois should not cause too much anxiety. In similar situations herbivorous populations confronted by a "new" predator have successfully developed defence mechanisms, while the most consistent damage has been suffered by non-endemic populations (mouflons in the Alps) or by red or roe deer, artificially accustomed to feeding from fodder troughs. We might also add that the reintroduction of the Abruzzo chamois to other Apennine massifs, starting from Maiella, is now under way;
- hostility from local circles - cultural, political, zootechnical or hunting, often considered elsewhere a serious obstacle to the lynx, does not seem to appear at present in Italy. In the Alps (Lagorai), the return of the cat has been welcomed. In the Apennines (Abruzzo National Park), a local authority has invested its own money to prepare an area to provide hospitality for the animal.

There are thus many reasons to consider seriously the return to our country, whether spontaneous or "sustained" by man, of this fascinating animal. The arguments in favour range from the beauty of a reintegrated living landscape to the rebuilding of the ecosystem all the way up to its most sophisticated components. It is known that the lynx, besides playing its part in effective natural selection and the containment of ungulate populations with predatory techniques quite different from those of the wolf and other carnivores, causes a beneficial dispersion effect on excessive and stagnant concentrations of herbivores, making the best use of food resources, preventing damage to vegetation and keeping the possible spread of epidemics at bay.

But perhaps the real reason the lynx is being recalled to our forests is even more profound, and stems from a conscious ecological drive for the restoration of the earth's values. Give back to nature, as far as possible, what has been unjustly taken from it, for its own good, which is also the good of mankind. That is why we say with great expectation and joy: Welcome home, lynx!'

It seems that despite waves of reports, Italy has yet to get caught up in any 'big cat' hysteria, the likes which have been sweeping the United Kingdom and the U.S.A. for decades. However, something strange does indeed prowl the hilly areas and rugged terra firma but until investigated with consistency, it's likely that the 'phantom' panthers will forever remain in the shadows of folklore.

MORE CRYPTIDS

Bigfoot, said to roam the Pacific Northwest of America, and the Yeti, of the Himalayan region, are probably cryptozoology's most famous names. Wildmen of Italy's rugged hills are certainly not household names and despite some of mountainous regions, tales of bipedal beasts are actually few and far between. Is this because such creatures are extremely elusive and whose territories melt into other countries, or because the belief in such hairy hominids is frail ? Rumour has it that in 1619 Italian philosopher Lucilio Vanini was burned alive for suggesting that man evolved from apes. I would have been interested to have heard his take on the modern reports of humanoid creatures unknown to science.

A bipedal gorilla-like humanoid was reported from the Sealza, in the Imperia Province in northern west

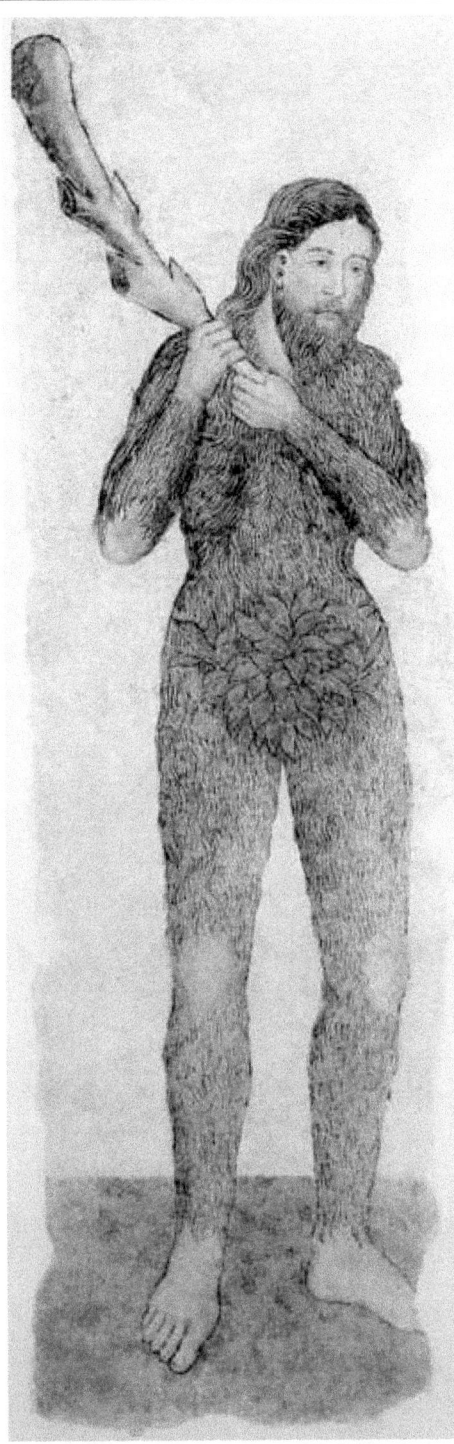

small, red creatures for some distance. Goodness knows what would have happened if they'd caught the terrified man, but Mr Sensi made it to hospital where he was prone to fits of panic which caused him to hide under the bed and fill the ward with screams. The following night at Parravicino d'Erba, 37-year old Renzo Pugina had just parked his vehicle in his garage when, standing in the shadows of a nearby tree, he observed a scaly humanoid which suddenly aimed a torch-like object at him and paralysed the witness from its beam. Renzo somehow broke the spell and bravely attacked the creature but the being hovered away, making a whirring sound. On October 27th a creature with ape-like eyes (!) but wearing a helmet appeared to an Ermellina Lanzillo outside her home at Grosseto. The humanoid was fat and had narrow shoulders yet paralysed the witness for a short time before she called her niece but the creature was gone. On November 2nd at Cremona, at 6:00 pm, two students out hunting encountered a weird dwarf which stood in a bluish cloud and had a rubber tube connected to its face which lead to a backpack of sorts. The witnesses ran in terror. Three freakish dwarf-like creatures appeared at Isola also in the same month to farmer Amerigo Lorenzini and on the night of May 19th 1960 a winged humanoid was seen at Siracusa by a man and his wife as they drove through town.

dwarf-life figure was seen in 1980 at Bisuschio, Varese, by a woman at 4:00 pm on the evening of February 11th. The being was concealed by trees as it seemed to skip from branch to branch whilst covering its face.

In 1986 at Padova, several witnesses had separate encounters with a reptilian humanoid that had glowing eyes. Despite the various reports details seem extremely vague except that the being was green in colour and very tall.

On the 16th December 1991 a student at a nearby university was startled to see a flying humanoid that resembled a robot in its movements but was covered in green hair and had glowing eyes.

Three years later at Rochetta a silvery-blue humanoid creature standing just fifty-centimetres in height was observed from a distance of fifty-feet

by an unnerved witness. The being then hovered and whizzed out of sight. In the November at Berceto a similar humanoid was observed, and also at Lentini

Website UFOINFO chronicled the following humanoid encounter,

'Location. Sacile, Pordenone, Italy
Date: February 27 1997
Time: 1700

Riccardo, while fishing alone by the River Livenza suddenly felt a presence behind him, turning around he saw a bizarre humanoid figure somewhat similar to a man. The humanoid was about 1.30 meters in height with an oval shaped head lacking a mouth, nose or ears. It had two large black oval shaped eyes covered with some form of retina similar to cobweb. Riccardo approached the humanoid and touched its skin. The humanoid's body was covered in a fine grey fur similar to that of an "astrakhan" rug. Its arms were disproportionately long as compared to the rest of the body and its hands were dark grey in colour with three long pointy digits that ended in pink coloured tips. Its feet were identical to the hands. The humanoid lacked elbows or knees and no sexual organs were apparent. It slowly approached Riccardo touching his forehead with one of its hands. At this point Riccardo felt as he had gone into some kind of psychic or telepathic contact with the humanoid. Suddenly a strange mist began to form around them and out of a mist a metallic landed object appeared. Riccardo was then taken inside the craft and shown around for about five minutes and then was deposited back on the riverside.'

There are many cases throughout the world involving bizarre creatures which also seem to have connections to the UFO enigma. Whether its creatures seen being dropped off by unidentified flying objects, or humanoids boarding such craft. Although I don't often like to connect the world of bizarre monsters to UFOs it's clear that some of these critters may be part of a bigger, more complex puzzle constructed of many parts. In areas such as Ancona and Turin, weird humanoid figures but wearing tight-fitting 'wet suits' have been observed in remote locations and although not always being seen near alleged craft, such beings have been seen to resemble the 'gray' alien beings which featured heavily in alien abduction reports across the world in the 1990s. Such humanoids have also been seen wearing strange helmets or carrying/using machinery such as lasers, metallic boxes etc.

In 2002 a luminous creature which scurried away from a motorist who was travelling at 1:00 am in the vicinity of Vigevano. The being was walking along the roadside but once the headlights picked it out the humanoid shuffled away. Five years later in Turin on a remote stretch of road, on 31st March at 10:30 pm, and man and his daughter travelling through Superga, witnessed an eighty-centimetre tall, whitish humanoid with big eyes that omitted a howling noise before running away into the night. In 2005 on April 28th at 11:30 am it was reported from Campania that a horrifying clawed monster had been witnessed walking the corridors of a local grammar school.

Thirty students became hysterical after observing the short, large-headed creature, and their statements were backed up by a teacher who also saw the being. The school chancellor advised the students not to talk about the incident to the press and the case faded into obscurity. Was the creature merely a delayed April fools prank? The previous year in Rome an Italian public servant claimed to see a strange thing flying on the sky whilst he was on his way home. Although he was unsure as to what he saw, for several nights the vision disturbed him and then on one particular night a peculiar humanoid creature appeared in his bedroom. The being was green and had huge black eyes.

It seems that a variety of truly monstrous and incomprehensible creatures roam some cold, desolate void within Italy's framework, but whether such unfathomable abominations originate from Earth or via some tear in the wall of time we may never know. A creature known as Frog lurks around Mantua as a devilish but frog-like humanoid, which is also given the name Fada. The monster is considered a bad omen, just like a handful of the spectral hellhounds are, which also prowl Italy's unknown corridors.

One of the scariest legends of Italy is that pertaining to the Lupo-Mannaro, a werewolf-type humanoid said to haunt parts of the Italian countryside, although legend states that such creatures form through men who were born on Christmas Night, and this story is strongly believed in the Naples and Campania areas. The Lupo-Mannaro is a lonely figure of folklore said to trudge through the rustic avenues during daylight hours and howl solemnly at the sky. In some remote areas the man-beast is feared so much that crucifixes line the dusty roads to repel his attacks.

Other legends which speak of the beast come from Sicily where the Lupo-Mannaro is said to be any child who was conceived during a full moon and in the Alps certain springs can induce bizarre transformations from man to beast. Versipellis, a Roman word translating as 'skin turner', are witch-like beings who are able to shape shift and often appear as hideous beasts.

The striges, or strix, are night owls related to vampyric attacks on children, and in Italy the belief in the vampire, or vampire-like monsters is strong also. Empusas, mormo, lemurs, lamia and larvae were all monstrous bloodsucking creatures which the Romans feared. Of course, such legends are far removed from cryptozoology but in my opinion should still be researched, as we can learn an awful lot from cultural fears and local superstition as many creatures which have been proven to be flesh and blood discoveries often began life as rumour and folklore.

Of course, in the case of so many colourful, frightening, surreal and confusing humanoids we may never know the dark truths behind their appearances. Their originations may be connected to some extraterrestrial void, or indeed the human psyche, and so I leave you with one last foray into the world of Italian monsters, and a creature known as the Crane Man, an anomalous humanoid sketched in 1642 by Ulysses Aldrovandi for his *Monstrorum Historia*. The entity was simply a human with a long neck and the beak of a crane on a human face. Such a horror appeared to a gullible European audience, just like all monsters do, but is it really a display of gullibility that enables such monsters into our realm, or the possibility that out there, somewhere, in the woods and waters of this world, there does indeed loiter a menagerie of the damned that no zoo, or scientific manual could ever accommodate ?

`JACK` THE HORSE RIPPER
Jon Downes

Dogs are popularly supposed to be a man's best friend. They have been domesticated for tens of thousands of years, indeed, it has been argued that it was in fact the dog that domesticated man rather than the other way around, because our ancestors used to follow the wolf packs around in order to feast on scraps. The second great friend of mankind is the horse. It was first domesticated about 10,000 years ago, and domestication has been widespread for the last two thousand years. Our species has been inextricably linked with dogs and horses ever since, and it is perhaps a sign of something spectacularly awry in the human psyche, that there have been so many reports over the years of vicious attacks on our two oldest friends.

I have been interested in the phenomenon of animal mutilation for many years. Indeed, on a number of cases I have been called in as a consultant by provincial police forces when faced with a spectacularly intransigent case. Mutilation is said to include removal of parts of the mouth and hind regions, especially the anus and sexual organs. Animals are usually said to have been drained of all or most of their blood, and some of the wounds inflicted are supposedly indicative of skilled or unusual procedures. Some attribute these mutilations to extraterrestrials because UFOs are sometimes reported in conjunction with alleged mutilations.

Horse-ripping, or horse slashing, however, is a peculiarly European phenomenon involving serious injuries in horses, often involving mutilation of their genitalia. Some people believe that these attacks are carried out deliberately by people, and generally sexually motivated, however, some are so bizarre that it is almost impossible to find either a cultprit or a motive. These cases are so intransigent that some people have even claimed that they can be explained by the animals harming themselves. This claim is partly based on the lack of people convicted for horse-slashing offences, and partly on cases where the injuries have been shown to be caused by other horses.

During the summer of 2003 I was contacted by a person I know who works for the RSPCA. Before we go any further I should say that partly because we have a strict code of confidentiality (and also because I want to protect my sources both within the RSPCA and the Devon and Cornwall Constabulary). My contact told me that there had been a particularly unpleasant, and intriguing, series of attacks on a horse in a little village about twelve miles south of Exeter. Was I interested?

As it happens, I was. Only the previous evening I had been reading about a classic case from the turn of the 20th Century, involving none other than Sir Arthur Conan-Doyle of Sherlock Holmes fame. George

Edalji, the solicitor son of an Indian-born vicar living in the wilds of Staffordshire was wrongly accused of 'ripping' horses, a gruesome crime involving taking a sickle to a horse and cutting off its genitalia, leaving it to bleed to death. George was serving a seven year jail sentence, when he contacted Conan-Doyle, and asked for his help. He was cleared, but the case was never satisfactorily put to bed; now it looked as if I had a chance to follow in the great man's footsteps.

It so happened that this attack took place in an area which was notorious for UFO activity. Over the years there had been a string of unexplained attacks on animals, and many people had linked them with the proliferation of UFO sightings and other strange phenomena. During the two or three days when the attack had taken place there had been a whole string of further UFO sightings, and one witness told us how he had been driving back along the deserted country road towards Exeter when he had seen a huge triangular object hovering in the air, some fifty feet above the road in front of his car.

My colleague emailed me a picture that had been taken by the attending RSPCA officer. It was horrific. To make this article easy-to-understand, when discussing the injury to the horse we use the words for human body parts, rather than those more usually used within the equestrian community. The main injury to the horse was a long straight wound to the right buttock. Approximately eight inches long, it appeared from first sight to be from a stab wound.

"Ratty", a 15 year-old gelding had been kept at the farm by a middle-aged couple. He had been a family pet and had been one of five horses owned by the family. The horse was fed and checked the horses at approximately 6pm on Tuesday 18th July, but whilst he left feed in the paddock for them on the following day, he did not check the horses individually until approximately 2 p.m. on Thursday 20th July. Unusually, the horses were all together in a small paddock next to the farmyard. They were usually kept in one of a number of other fields across the property. In the corner of the paddock is a small outbuilding. The owner entered the outbuilding and found Ratty lying on the floor. Upon investigation he found that the animal had been injured. The vet was called to the scene and found that the wound itself was not that deep, indicating that it was a stab wound from an implement like a Stanley knife or craft knife, that had become infected.

There were number of other wounds on the horse's body. There was another long slash mark on the opposite buttock. This was not deep enough to cut the skin but left a very noticeable scar. There were other wounds on the flank and legs. Several of these were deep enough to have drawn blood. There were many other strange marks, many of which appeared to be from older injuries to the horse across his flanks legs and ankles. The marks on the horse's ankles were particularly interesting because if they appeared to be bite marks inflicted by some small carnivore. On the flank there was an old scar, which, we were told, had been inflicted the year before when Ratty had apparently been attacked by an unknown predator, possibly a big cat.

We found out that there had been attacks on animals at two different farms in the district during the period when Ratty had been attacked. This suggested that any thought of a personal motive for the attack should be discarded. It also transpired that that had been attacks at roughly the same time of the previous year, and we are minded to suggest that the attack on Ratty the previous year was not by a big cat but was almost certainly by a human. The fact that the attacks had taken place nearly a year apart and - in Ratty's case - on the same horse suggests to us that the same culprit was responsible on both occasions.

By this time we were working closely with the police, and although not entirely convinced that the attacks could have been carried out by a human-being, we gave the Devon Constabulary the best human interpretation of events that we could manage. The owners were convinced that the attack had been by someone with a grudge against them, but I believe that to be unlikely. Grudge attacks often intend either

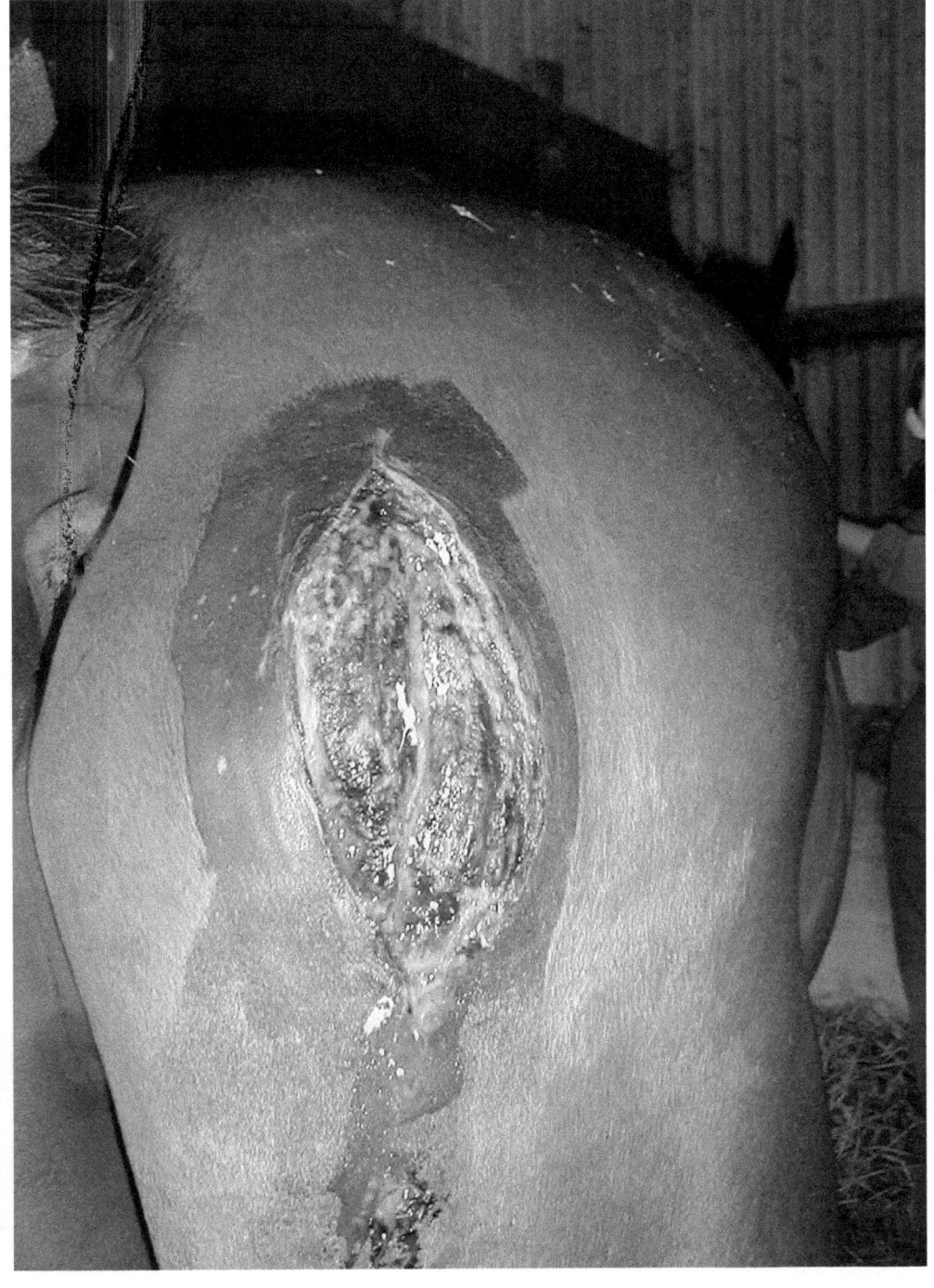

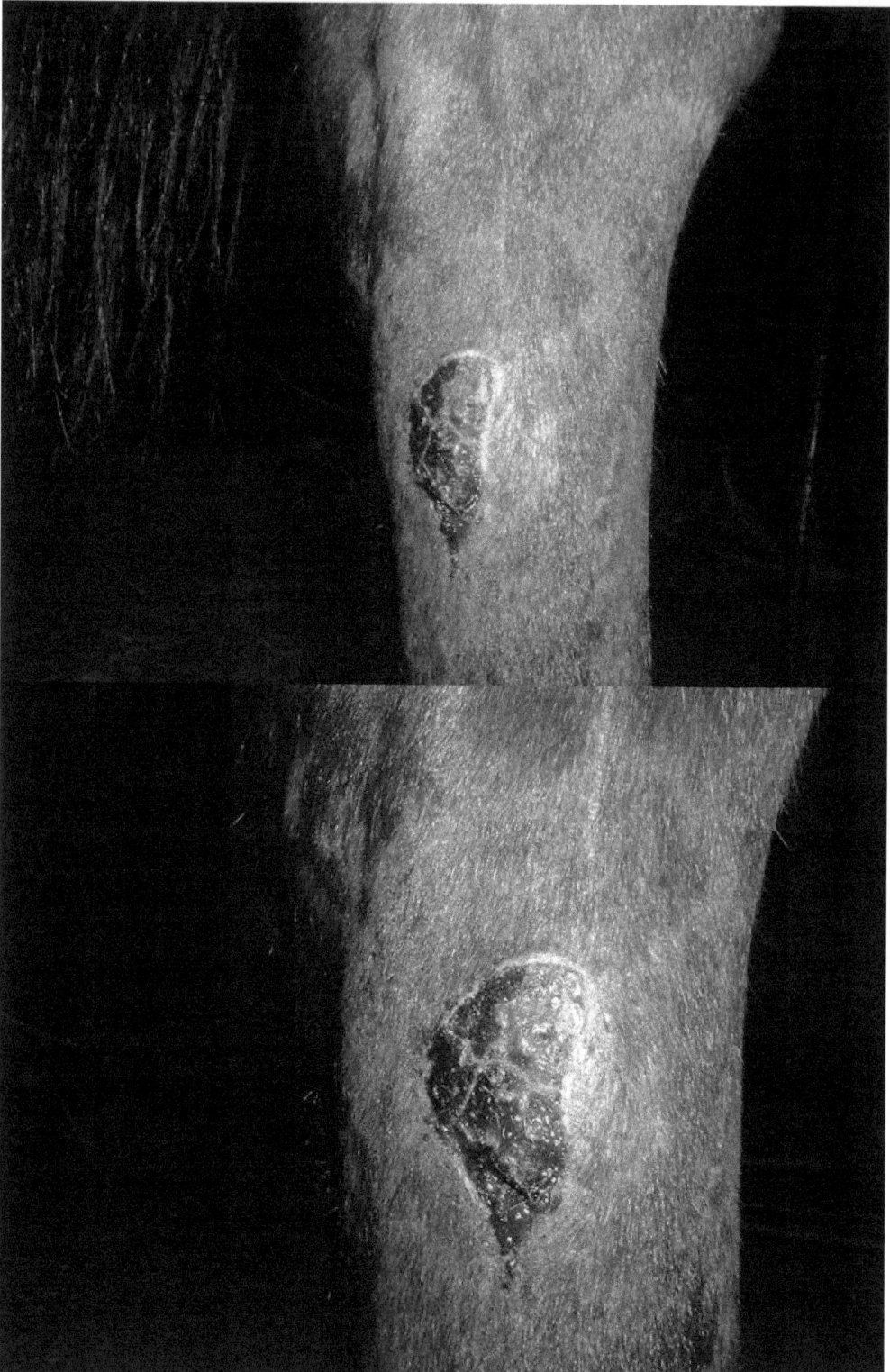

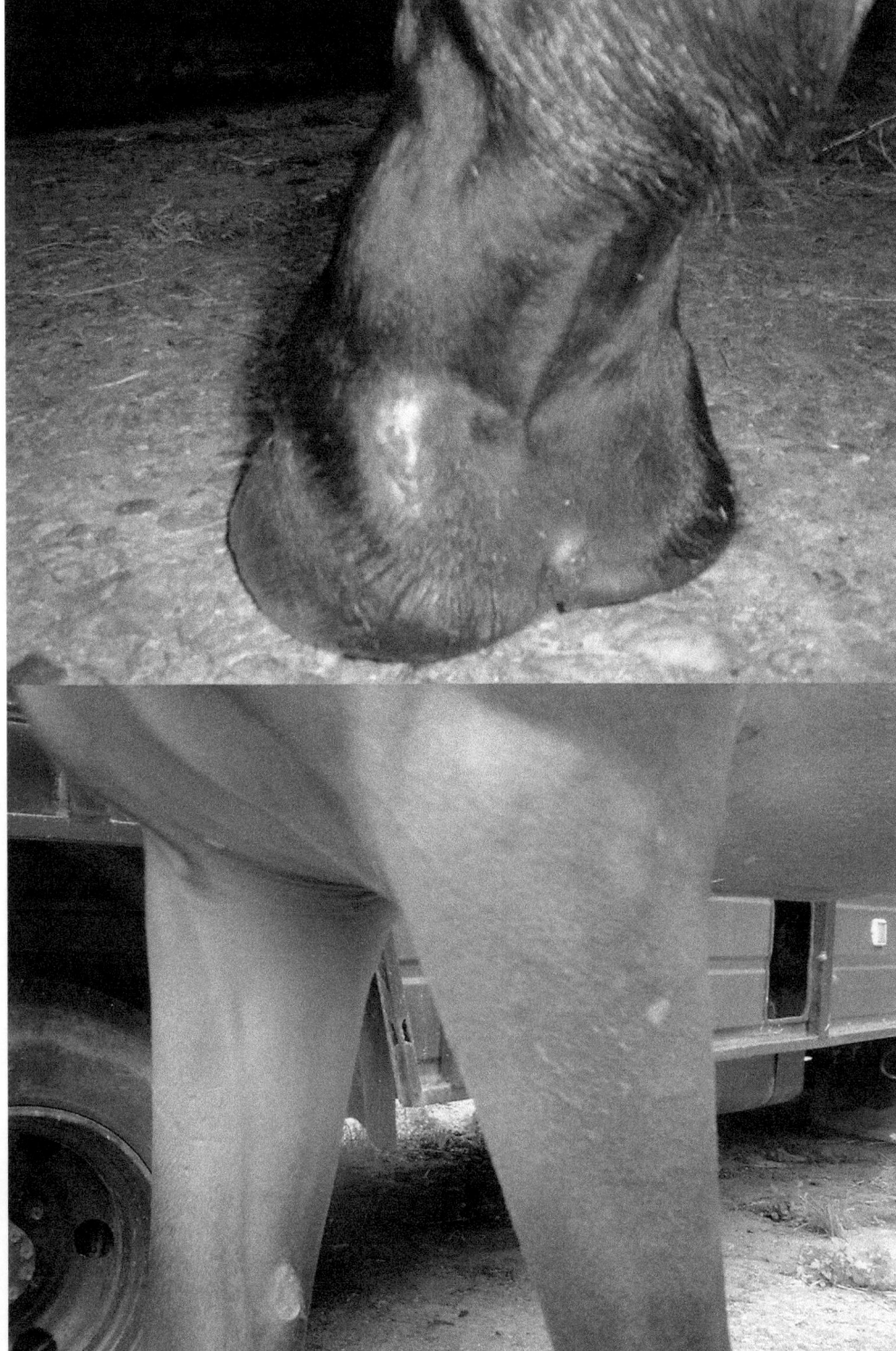

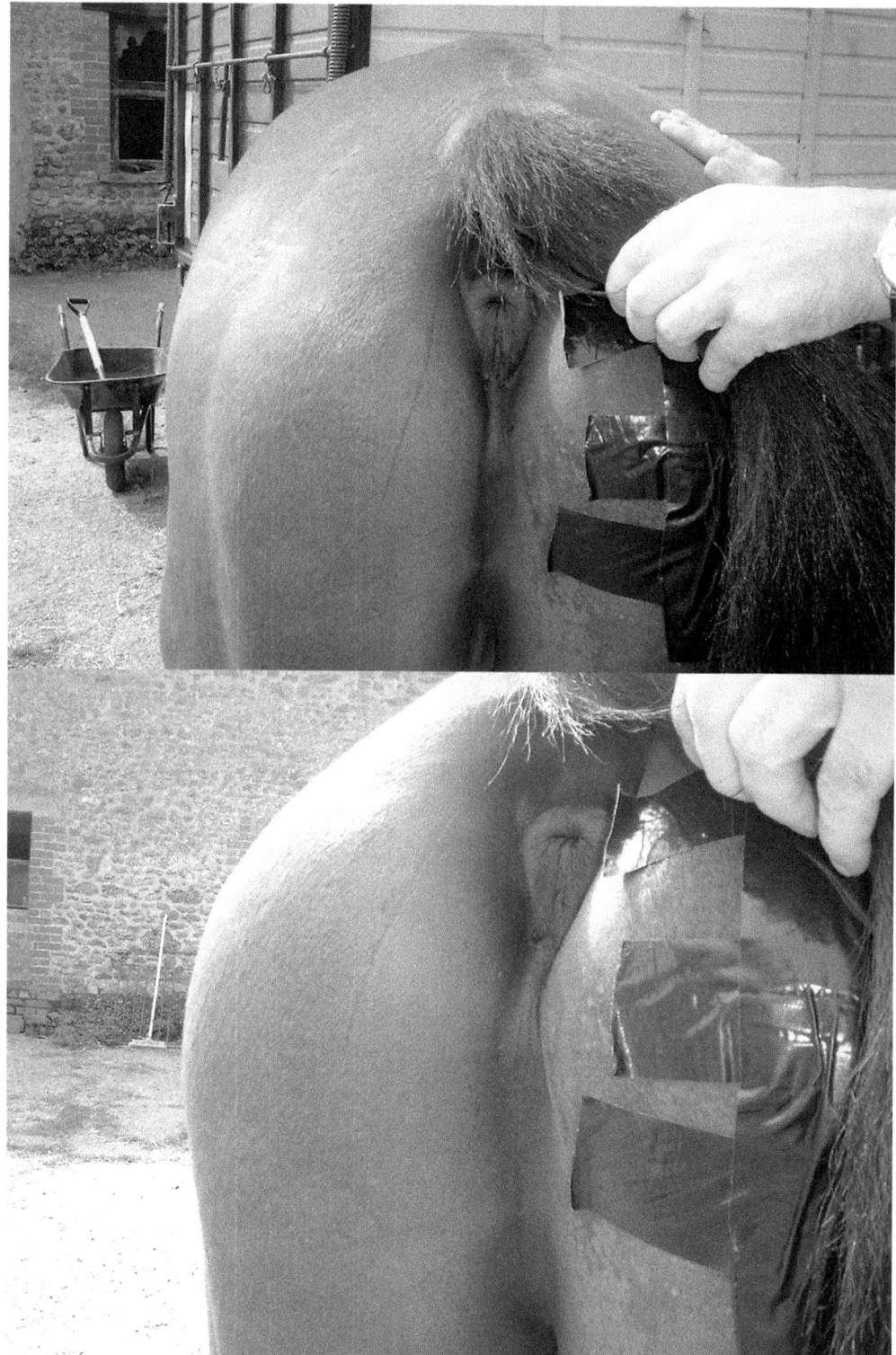

to kill or to permanently disfigure, and it is these allegations are when ears are have been severed, eyes gouged, manes burnt, and in the most extreme cases decapitations carried out. So, if the attacker were, indeed human, what other motive could there be?

A colleague of ours, a computer forensics expert who works with the police from many forces and is also a keen horseman. He told us that in recent months there had been a minor explosion in the volume of horse related pornography. This was borne out by several other people. It was then that we realised the significance of the scratch mark on his other buttock. The image that have reproduced here, is, we believe, evidence that the perpetrator or was not trying to maim Ratty. He was not even trying to draw blood. He was merely trying to impose his mark upon the horse in much the same way as a graffiti artist takes pride in putting his 'tag' on a moving vehicle. The perfect symmetry of the two cuts is, we believe, the key to the case. The first cut - on the left buttock was successful, the one on the right-hand buttock completely unsuccessful. We believe that the motivation was to make these two marks and then take a photograph which would be the perfect piece of horse pornography. In the event the horse got badly injured, and ran up panicking through the brambles and undergrowth which caused the vast majority of the rest of his injuries. The perpetrator made his escape probably without even taking the photograph he so badly wanted.

The fact that Ratty and other horses in the area had been attacked a year before, but got no subsequent attacks - had been reported led us to hypothesise that if there is a person responsible, it is a visitor who only comes down to Devon for a week or two in the summer. Whereas most people visiting the area for their summer break look forward to a visit to the seaside, or Devonshire cream tea, our unknown assailant quite probably spends his whole year looking forward to getting some more pictures of mutilated horses on his digital camera.

But that doesn't really explain anything. How did the attacker get into the farm in broad daylight, without either appearing on security cameras or disturbing the three highly noisy guard dogs? How did he get away again? Although it is possible to access the farm across neighbouring fields, to have done so in the daytime without being seen beggars belief. Devon is full of stories of unexplained attacks on livestock, some in conjunction with UFO activity, some which appears to be ritual in nature, and some which are even linked with stories of werewolves which proliferate across the county and have done for centuries.

The unpleasant truth is that although the police were content with the solution we proposed, no-one has ever been charged with the crime, which remains unsolved, and perennially intriguing.

TELEGRAM FROM OZ: 2008 REPORT OF THE AUSTRALIAN BRANCH OF THE CFZ
Ruby Lang

The past 12 months have flown by, and due to our dedication to one project in particular – our research into big cat sightings and predation across Australia - we don't feel we have done a heck of a lot else.

The media have certainly followed the story of the big cats closely, and it would be fair to say we've been lucky enough to enjoy some of the best, balanced stories that have ever been done on the enigma in Australia.

Two reports in particular were screened by current affairs shows – programmes that usually do not rate highly on the credibility scale, and tend to focus on tales of dodgy landlords, suspect tradesmen and fathers who shirk child support duties – that opened the floodgates to many reports.

Disappointingly, despite seeking our help and shamelessly plundering what we had, we did discover that working with some TV folk is not a two-way street.

One nameless programme boasted of receiving 150 emailed reports after our interview, but declined to share them based on concerns about contravening The Privacy Act. Those kinds of responses can be quite frustrating, and you can be certain that the program will do nothing with them – they're probably already in the virtual bin!
But we digress.

Our first task in opening up a CFZ presence in Australia was to set up a parochial Australia CFZ presence on the web, which we have done in the form of a blog at http://cfzaustralia.blogspot.com/.

The talented Mark North obliged us with a customised CFZ Oz logo, with a great Australian flag offsetting our national mystery animal, the Tasmanian Tiger.
There are now plans afoot to set up a national CFZ phone hotline.

We feel that a dedicated service, if run even for a short period of time, may encourage the reporting of sightings of mystery animals.

We'll keep members abreast of developments.

Several months ago the CFZ Oz email list, a who's who of Australian cryptozoology, took on a life of its own and sparked the sharing of reports on an unprecedented level. It was originally set up to alert people to new posts on the CFZ Oz blog, and share news articles of interest.

While not strictly 100% Australian anymore – we have a New Zealander and a few CFZers from the UK and Scotland – the list makes for interesting, if somewhat sporadic, reading. It's nice to see that the competitive nature that often creeps into fringe research is missing here, and the camaraderie quite strong.
Of course, the big project Down Under that has been absorbing all of our time is our new book, which we are just putting the finishing touches on – *Australian Big Cats: An Unnatural History*.

It promises to be the big cat book to end all big cat books in Australia (with a tip of the hat to the late David O'Reilly, whose *Savage Shadow* remains a cult favourite, and authors Paul Cropper and Tony Healy, whose *Out of the Shadows* also does a thorough job of exploring the ABC enigma), and will make gripping reading.

We have amassed an incredible archive of photographs and video footage, as well as countless sighting and predation reports, input from scientists and excerpts from many 'secret' government reports.

And in the course of researching the book, we had an interesting brush with the law – we uncovered a fake zookeeper operating in rural Australia.

You can read about that incredible 'inside story' in issue 45 of *Animals & Men*. It's rarely dull Down Under!

In addition to the media work we've carried out – always, we must stress, with CFZ t-shirts at the ready – there has been several excursions interstate.

Most notable was a trip to South Australia where Mike Williams tested out a forward-looking infrared camera (FLIR), and chased reports of rampant predation and sightings of large black cats with CFZ friend Darren Croese.

Although costly (about $400 for three days) the camera has proved its worth as recommended field equipment, and we hope to deploy it again in the field soon. There's no hope of buying one at this stage – brand new models start at $10,000, and seeing as we often scratch around for $10 to help make up the cost of a pizza, it's unlikely we'll have one sitting on our shelf anytime soon.

Other trips to Victoria and around NSW have yielded further reports, and in one case solved an interesting cryptozoological mystery – but you can read all about it when our book is out.

That's all from Australia. Cheerio!

CFZ-USA 2008:
A Cryptozoological Round-Up
Nick Redfern

When I moved to live in the United States in 2001, one of the things that your esteemed editor and I realised very early on was that my change in both location and circumstances presented a great opportunity to really push and promote the work of the CFZ in pastures distinctly new and different.

To an extent at least, my column – *Nick Redfern's Letter from America* - that regularly appears within the pages of the CFZ's in-house magazine, *Animals & Men*, is indeed designed to keep interested parties informed of what's going on with the good-ship CFZ-USA.

But there's only so much that I can relate in a couple of pages of a quarterly magazine. And with that issue in mind, just recently Jon asked me if I would be interested in submitting to the pages of this *Yearbook* a summary-report that would detail some of the work of the CFZ-USA Office in 2008.

Well, of course, my answer was an immediate 'Yes.'

But before I get to the meat of the matter, however: a few words about the running of a body like the CFZ-USA. And, specifically, this is intended for those of you that may be contemplating approaching Jon with a view to setting up your own off-shoot of the main Woolsery, Devonshire-based CFZ office.

It always amazes me how so many people within Forteana seem to assume that if you want to promote your group, your book, or whatever, that the publicity will come running to you. In reality, however, it rarely works like that. One of the things that I've seen lacking time and again – and it's something that infuriates me – is the inability or gumption on the part of certain Fortean players to forge links, and I mean deep links, with (a) local media outlets, such as newspapers, radio, and TV; (b) nearby bookstores and libraries; and (c) local groups that may want to hear more about your work.

The harsh fact is that without such contacts, many groups – no matter how noble or laudable their plans are – simply don't survive.

Now, I want to stress that there is nothing egotistical about this. If you want to survive and thrive in Forteana and cryptozoology as an individual or as a group, then doing investigations alone is not enough.

Of course, investigations are vital when it comes to trying to resolve what lies at the heart of some of the mysteries we are all pursuing. But a strong presence at a local level, the ability to deliver the goods when the media needs them, and always ensuring that that same media is kept fully up-to-date on what you are doing, are all vital matters; the importance of which cannot be stressed enough.

Don't get me wrong: manning the CFZ-USA is great fun and massively entertaining, but it is damned hard work, too – as Jon will be only too keen to tell you from his own perspective in jolly old England. It requires diligence and patience; and it requires the ability (and the willingness) to spend hours on the phone doing interviews, chasing down leads, promoting the group, and much more.

Having just gone through my day-planner for 2008 thus far, I can tell you that, today, the CFZ-USA averages seven radio interviews per week, three lectures per month, one conference per month, and a mention in three magazines and/or newspapers per month too. And, as a result, we are making great headway in terms of getting the word out about what we do, and, equally importantly, uncovering new and important data and cases.

So, if you too decide to take a tentative step into the exciting, adventurous, murky and at times hazardous waters that go with running a branch of the CFZ, I urge you to embrace your dream and follow your heart. But be aware that it is about far more than merely searching for monsters. It's about taking a professional approach, ensuring the group's survival, and making the most of the media whenever and wherever possible.

And with that all now said, on with the story, and a few of the things that have been happening this year in the world of the CFZ-USA.

Monsters of the Woods

On the evening of January 15, I gave a lecture on my *Memoirs of a Monster Hunter* book for the Texas-based *Denton Area Paranormal Society, DAPS*, organised and run by a man named Lance Oliver. The lecture went very well, and during the course of the evening Lance happened to mention that he had undertaken investigations of a couple of local stories, legends and incidents of a monstrous nature that he felt might very well be of keen interest to me.

And indeed they were: one concerned repeated encounters of a classic 'Goat-Man'-kind at an old bridge only a few miles from the lecture-hall; and the other involved a series of Bigfoot sightings at a large body of water nearby called Ray Roberts Lake. Needless to say, a first-hand trip to the locations in question was warranted. Thus it was that a fortnight later, Lance and I headed off in search of the unknown.

As we arrived at the Old Alton Bridge – the alleged hang-out of the Goat-Man, and which was constructed back in 1884 – Lance told me the story, or rather the stories, behind the legend. One suggested that back in the 1960s occultists in the area had inadvertently opened a portal that allowed the unholy creature to enter our world unhindered. As a result, he continued, the hellish Goat-Man had now made its new home within the deep waters and thick woods that surrounded the old bridge.

The second story was even weirder. According to legend, decades ago a local man killed his parents and was hung for the crime. At the moment of his death, so the tall-tale went at least, the man's head was torn from his neck. As a result, his ghostly frame returned from the grave looking for a new head – which was duly acquired late one night from a wild goat that was unfortunately wandering around the Old Alton Bridge at the time. Of course, this was surely nothing more than a wild tale spread by local teenagers and thrill-seekers. But, for the CFZ-USA, it was yet another case to add to an ever-burgeoning series of files.

Next stop: Ray Roberts Lake – which, in times past, was the haunt of various Native American Indian tribes, including the Comanche, the Kiowa and the Tonkawa. It was at Ray Roberts Lake that, in May 1990, strange animalistic screams were heard late at night, and on more than one occasion a giant, hairy man-beast was seen on a large hill on the north-side of the lake.

As I have come to learn, in areas where Bigfoot has been seen on regular occasions, it is not uncommon to find what have become known in the world of monster-hunting as 'Bigfoot Teepees'. These curious structures are basically made of thick branches that look like they have quite literally been wrenched off trees and placed into pyramid-like formations. Some researchers have suggested they are made by the creatures as territorial markers; while others have postulated they might very well be connected to Bigfoot mating rituals.

A definitive answer to their construction still eludes us; however, I do not exaggerate when I say that the peak of the wooded hill was absolutely infested with such formations. To the left of us, to the right of us, in front of us and behind us: they were everywhere. Of course, out came the cameras to preserve the astounding scene.

The whole place reeked of menace and high-strangeness; and I got a distinct and unsettling feeling that hidden eyes (and probably glowing red eyes, too…) were watching our every move. This was not a good place; not at all. But it was one that I would not forget as I sought to add to the CFZ-USA's ever-burgeoning data-base of material.

Supernatural Big-Cats

Following a profile on the work of the CFZ-USA that appeared in a Dallas magazine in May 2008, I was contacted by a man named Danny Shaw who lived only about forty miles away, and who had a controversial story to tell. Granted, it was a story that, in some form or other, I had heard before. But that didn't make it any the less fascinating.

The essence of the story related by Shaw was that while driving through woods near the Texas town of Decatur late one night in February 2008, he had been startled by the sight of what he described as 'a large black-panther running across the road' in front of him.

Well, the idea that there might well be big-cats on the loose in the wilds of Texas is not that strange, you might reasonably think. There was something very different about Shaw's report, however. As I sat in his living-room a week or so after he initially contacted me, Shaw said that although the beast looked like a big-cat, there was one aspect to the incident that had left him both flummoxed and concerned.

When I asked him what he meant, Shaw replied that the creature suddenly vanished – and I do mean literally vanished – as it was about half-way across the road. When I asked him to clarify what he meant, Shaw replied firmly that it had not disappeared into the shadows or into the safety and camouflage of the trees; but 'vanished like it was invisible'.

Somewhat embarrassed and acutely aware of the controversial nature of his story, Shaw asked me for my opinions. I told him that although most big-cat researchers were far more than loathe to bring the paranormal into the puzzle (and particularly so in Britain), the fact was that there were indeed a number of big-cat reports that were suggestive of very high-strangeness being in evidence, rather than these things merely being exotic escapees from private-zoos, and such like.

Interestingly, I learned further from Shaw that the stretch of road on which the cat both appeared and vanished was near an old cemetery which had had far more than its fair share of uncanny activity over the years – including sightings of both giant wolves and a large, writhing snake. And the fact that there were said to be 'devil-worshippers' in the area, too, only made the story even stranger and much, much darker.

Most significantly of all, Shaw was able to refer me to three other people who had seen spectral big-cats in the same area; and I am now in the process of preparing a paper for future publication titled *The Phantom Big-Cats of Decatur*, which will be a major CFZ-USA report on this particularly perplexing aspect of big-cat studies.

The North-East Cryptid Initiative

In June, I was very pleased to bring on-board our latest U.S. representative: Brian Gaugler of New York. Brian is covering his native New York, New Jersey, and Connecticut, and is already hard at work digging deep into old cases, new ones, undertaking research, filing case reports for CFZ-USA, and working on papers for future publication.

Brian is also in the process of establishing the *North-East Cryptid Initiative*, an exciting project, the aim of which is to create a solid network of researchers who can share data, and stay informed of what is afoot in their respective areas.

Brian is particularly keen to hear from researchers in the North-East sector of the United States, and encourages anyone who may be able to help to contact him at: brian.gaugler@yahoo.com

And as he says:

'Basically, what I hope to achieve with the *NCI* is to establish a network of researchers and groups within the North Eastern region of the country who can share information and data with each other and work together more effectively and efficiently in moving the field forward and undertaking, and promoting, quality work concerning mystery animals in the North East.

'I strongly feel that this part of the country is often overlooked by the cryptozoology field on a national level, and thus a great deal of solid, quality data never receives the attention it deserves, and a lot of credible, serious researchers never receive the recognition they deserve, either.

'Therefore, by creating a united effort, I truly think that all involved can help make the regional into the national, or take the information that, up until this point has only been known at a local level, and make it more well known at the national level, so that researchers all over the country are aware of crypto reports from Maine or New Jersey, for example. I also think, by bringing people together and getting them talking with each other more, it can help to overcome the arguing and strife that has taken over a large part of the cryptozoology field.

'I believe this to be one of the biggest problems facing the field as a whole, the clashing of personalities and the constant squabbling, when all involved should ideally be unified to pursue their common interest. Thus, I hope that the *NCI* network will at least be a step forward in solving this problem, as well as helping researchers become involved who wish to break into the field.

'The next step I wish to take in order to get the *NCI* fully established and up and running is to get a web presence set up. I think that a blog site would be the best choice, since it is free and easy to manage, so if anyone is interested in helping me to set that up, please let me know.

'Also, I wish to have a few articles/case reports in hand in order to put them up immediately once the blog is set up, so if anyone can help me out with that, again, please contact me and let me know. I am also going to start working on a list of guidelines when I get the chance, I welcome all suggestions and advice on that or anything else, please feel free to voice your opinions/advice/criticism, I want this to be as democratic as possible.

'Please don't hesitate to contact me, and thanks everyone. Let's get rocking and rolling!'

Needless to say, as Brian is a key CFZ-USA representative, he has our full support as he forges ahead with the *North-East Cryptid Initiative*. After all, greater cooperation, an increased sharing of data and a less self-serving approach to the subject of cryptozoology are all vital if we are to push forward and actually have some form of resolution, rather than merely accumulating more unsolved cases, ad infinitum.

Conversation with the Blogsquatcher

As the summer of 2008 was at its height, I was interviewed extensively by the *Blogsquatcher* – who has an interesting, and cutting-edge blog on the Net that delves into distinctly controversial areas that others prefer to shy away from. And so, when he asked me if I was willing to be interviewed about my thoughts on Bigfoot in Britain and the attendant high-strangeness that accompanies many such reports, I was quick to respond in the affirmative.

In the interview I stated:

> '...several things stand out when you talk about the British Bigfoot. One is that in many of the reports, the areas in which they've been seen [have] some sort of significance. Let me explain what I mean by that. A lot of Bigfoot sightings in Britain have been made in the vicinity of ancient stone circles, prehistoric burial mounds, and areas that ancient man perceived as being of some significance. Whether magical, cultural significance, whatever. One classic example is a place, ironically not far from where I was born and grew up, called the Cannock Chase.

> '...There's an area on the Cannock Chase called the Castle Ring. Now, the Castle Ring is an ancient structure built by ancient man, I think something like four or five thousand years ago. There are all sorts of theories concerning why it was built, [and] what its significance was and is. But the interesting thing about castle ring is that it has been the sight of nine or ten Bigfoot-type encounters that I'm aware of at least.

> 'This includes people seeing shadowy forms in the woods, bright red eyes I actually had a very weird experience myself once over there when staking the area out, and saw what looked like a flitting, shadowy form with these two red eyes. Now of course people can say, "You're seeing things that you're looking for," and what are the chances of something being in the area at the time you're looking for it. I would agree with those sentiments and those statements. And that's why I think the British Bigfoot, at least, is different in some ways to some of the reports from elsewhere. It is the case that they, you know, turn up in these prehistoric locations, and sometimes people are looking for them. It's almost as if these creatures know and manifest for them. Now I know to some people, they think, "Oh, Nick's off on his paranormal tangent again," and there's like a little rolling of eyes and shaking of heads...'

It was this particular interview that perhaps generated more comments and questions with respect to the work of the CFZ-USA, than any other I've done in a long time – and I most definitely recommend you check-out the *Blogsquatcher* for a wealth of thought-provoking and stimulating debate.

A Close Encounter of the Chupa Kind

One of the most fascinating stories to reach the CFZ-USA, and that I hope to make a personal, on-site study of in 2009, came from a man named Mac, who submitted the following account:

'I find the possibility of the Chupacabras particularly interesting, as it is the crytpid I may have gotten a brief glance at. During the Florida drought of 2001 I lived on a farm with my ex-wife. Many of the trees were in great distress because of the heat and dry conditions. One had nearly fallen over on my house. In an effort to help cool and water them I was out spraying down their trunks during the hottest part of the day. I was using a pressure nozzle with some real power.

'At one point the stream went into an open cavity and out popped two very unhappy looking creatures the like of which I have never seen before. Rather large, especially given the size of the hole they emerged from, about three feet long, they gave much the appearance of a primate and moved like one. Their shoulder [sic] looked strong, even bulky. They had flat faces, and I remember they seemed to be squinting against the light.

'Most curious of all, from their arms to their legs stretched a thick membrane much like a bat. They were startlingly white. It could be said that these were just a large albino bat; in and of itself that would be quite a sighting. However, the largest bat in North America is called The Western Mastiff bat which in the U.S. is only found in southern California, and the body of which is only a foot and a half long.

'Honestly, as someone who has studied wildlife science, the size of the wings doesn't seem large enough to carry a creature of that size. Is that a Chupacabras? I don't know. But it was something. It was not an Opossum, as there was no gray in the fur, no naked tail, and it moved completely differently. The sighting didn't last long. I remember feeling bad for them actually, as though I had disturbed their privacy. I got the impression they were either siblings or a mated pair. They gave off no sense of menace or evil. Strangely, I did feel as though they were sentient somehow, different than just an animal, and their heads were quite large, with the rounded, side mounted ears of a primate. It's just strange. I looked for them after that, but never saw them again.'

A Monster Lecture

Midway through 2008, I was contacted by a man named Christopher Bader, who is an Associate Professor in the Department of Sociology at Baylor University, Waco, Texas. It transpires that Chris has a keen interest in cryptozoology. As a result, he invited me to speak for his students on the subject of my research, my investigations and the work of the CFZ-USA. And which is precisely what I did on the late-morning and early-afternoon of September 25.

Having first detailed the history and origin of the CFZ, I dug deep into some of the data in my book, *Man-Monkey: In Search of the British Bigfoot*, as well as such matters issues as Mokele mbembe; werewolves; lake-monsters; accounts of still-living pterosaurs; the Chupacabras; and much more of a definitively monstrous nature.

Normally, I lecture to audiences who are already deeply acquainted (or at least fairly acquainted) with the subject matter in hand; however, this was a slight departure, in the sense that those in attendance -

approximately 130 in number - were students at the university, many of who had little, or no, previous exposure to the world of cryptozoology.

The day turned out to be a very good one, however, with a wide variety of questions asked both during the lecture and in the immediate aftermath. I also detected more than a passing interest in some of the stranger, paranormal-style aspects of cryptozoology - such as those specifically exhibited by the British Bigfoot (including its ability to seemingly appear and vanish at will, and its ubiquitous, glowing red-eyes), which I thought was intriguing. And, of course, it was the perfect opportunity to promote the work of the CFZ-USA, too.

Monster-Eels

'I am British and live in Florida. My family and I came to Florida by sea from Australia in 1969,' wrote John Weatherly to the CFZ-USA. 'Our ship left Acapulco and sailed along the west coast towards the Panama Canal. It was the first week of July 1969. The sea was calm and we were cruising quite slowly because of congestion in the canal. As we cruised along the west coast of Costa Rica and Panama we were about 7 or 8 miles from shore and just a few yards from the flotsam line. It was clearly defined line of sea weed about 30 feet wide with odd bits of wood and the occasional small tree limb.

'We cruised along this path for several hours in bright sunshine between about 10 AM and 2 PM. There were many fish visible and some very large turtles but the significant sighting was huge eels. These creatures were always in pairs and we saw a pair perhaps every 20 minutes or so. They averaged about fifteen feet long and had a diameter of about one-and-a-half feet.

'They were khaki or olive in color and were identical to the eels for which I used to fish as a boy in my home town of Canterbury Kent, except they were so large. They were lazily swimming very slowly along through the flotsam or just wallowing at the very surface. The ship was carrying about 1,200 passengers and most were on deck on this idyllic day so the eels were seen by many people. Most were engaged in counting the enormous numbers of sharks which were clearly visible around the ship.

'I wonder if you have any idea what species of eel these were? They could easily have swallowed a child or a small adult.'

I wondered, too, and I duly sent John a bunch of photos of eels known to inhabit the waters in question, but he advised me that the beasts he saw were far stranger in nature:

'Thanks for the inputs, Nick. Unfortunately they did not add a great deal to the identity of the eels that I and my fellow passengers saw all those years ago. The images are still very clear in my minds eye. I am not much of an ichthyologist or zoologist for that matter, being a retired communications engineer by profession, so I can only speculate and very possibly be in error. However, I would suggest that since we only saw these creatures in pairs that possibly they had come to the surface for mating as I believe do some other relatively deep water species?

'Also, every moray eel picture that I have ever seen usually depicted a fish with a spotted or patterned body. These were not like that but a uniform smooth light khaki or even green/mustard color. Also the snout of the moray is quite pronounced. If memory serves me correctly the ones we saw had a more rounded nose. I do not recall the eyes being specifically prominent either although we are now stretching my memory a bit. I only wish I had access to a good telescopic lens for my camera at the time.'

Having heard John's story, so did I.

Whatever the true identity of the giant eels of the Panama Canal, it seems they are destined to remain a mystery.

Next Year

2009 promises much more. Fellow Texas-based monster-hunter Ken Gerhard and I will be embarking on the writing of a book on mystery animals of the Lone Star State; I am looking at the idea of setting up a company to provide guided-tours to those that want to visit some of the more notable locations where cryptids have been seen in the United States; and I will be vigorously promoting the work of both the British and American Offices of the CFZ at the following events: (a) *The Jefferson, Texas Paracon III Conference* - March 21, 2009; (b) *The Beyond Reality Conference*, New Hampshire - April 24-27, 2009; (c) *The Haunted America Conference*, Decatur, Illinois - June 19-20, 2009; and (d) *The Mysteries of the Universe Conference*, Kansas City - July 16-18, 2009. And, if you can make it to any of these gigs, say hello!

Nick Redfern can be contacted at his website: nickredfern.com

CENTRE FOR FORTEAN ZOOLOGY ANNUAL REPORT 2008

Dear Friends,

I can hardly believe that this is the fourteenth time that I have sat down in the first week of December to write my annual report. A heck of a lot has changed, both for the CFZ and for me personally, since I first sat down to write an account of what the CFZ had achieved in the previous year.

We have gone from being a small, benign, and – if I am to be completely honest – pretty ineffectual organisation of about 100 people, to being the world's largest cryptozoological research organisation, with over 400 members, and hundreds more sympathisers across the world. We have the world's largest dedicated fortean zoological publishing house CFZ Press, and our own multi-media website CFZtv.

OVERVIEW

In the past year we have published nineteen books, with two more – the 2009 Yearbook, and Neil Arnold's *Mystery Animals of Kent* - due before the end of the year. We have published four issues of *Exotic Pets*, and three issues of *Animals & Men*. We have produced 12 editions of *On the Track*, our monthly webTV show, ranging between 18 and 32 minutes in length, and a feature-length documentary, with two others in the pipeline. We promoted our ninth annual conference, and had the entire thing - all seventeen hours of it - available on CFZtv within a week. We also helped present the 2008 Big Cats in Britain Conference and presented that in its entirety on CFZtv. We finished building work on our museum, and carried out a three-week expedition to southern Russia, within weeks of the entire area becoming a war zone.

At the risk of sounding too self-congratulatory, I think that we have done quite well this year.

YEARBOOKS

The year started with personal tragedy as my father-in-law died at Christmas, and much of January was spent helping tie up the loose ends of his estate. We then threw ourselves into the tortuous exercise of republishing the CFZ Yearbooks from between 1996 and 2003. I am ashamed to say that some of them were a disgrace, with typographical and formatting errors beyond our wildest dreams. However, in our defence, I must say that we were one of the few organisations in the world, who – with a core team of just three people, and from 2001 a part time office assistant – would have embarked on such an ambitious exercise with no funding. And also with no way of publishing them apart from a photocopier and a comb binding machine. I am not sure whether we were being brave, or just stupid! However, the process of reformatting them for the 21st Century, and upgrading them to the professional standards of production that is the norm for us these days, has been a remarkably complicated job. The first three – 1996, 1997 and 1998 – have been presented as facsimile editions, with most of the typographical and print errors removed, and the images updated where possible, but the others (1999, 2000-1, 2002, 2003, and 2004) were re-typeset from scratch.

This was a horrible job, and was not finished until the end of May. In the meantime, we are proud to have published the 2008 Yearbook, the Guyana Expedition Report, another volume of collected editions of *Animals & Men*, a remarkable book by Michael Woodley - a remarkable young man, of whom more later - and the 2008 Big Cats in Britain Yearbook. This was the third of these volumes that was produced for them by us. However, the BCIB group have decided to produce the 2009 volume themselves, which is – of course – their prerogative, and we wish them well with this and other future endeavours.

MYSTERY ANIMALS OF THE BRITISH ISLES

In June we launched a new series of books - 'The Mystery Animals of the British Isles'. The first book in the series was by our old friend Mike Hallowell, and covers Northumberland and Tyneside. We eventually hope to cover the whole of the British Isles, including Ireland. However, in order to protect delicate political sensibilities, the volumes covering both Eire and Northern Ireland will probably be titled 'The Mystery Animals of Ireland'.

Future volumes planned include:

The Mystery Animals of the British Isles: Kent by Neil Arnold (due December 2008)
The Mystery Animals of the British Isles: Co. Durham and Teeside by Mike Hallowell (2009)
The Mystery Animals of the British Isles: Staffordshire by Nick Redfern (2009)
The Mystery Animals of the British Isles: Greater London by Neil Arnold (2009)
The Mystery Animals of the British Isles: Western Isles by Glen Vaudrey (2009)
The Mystery Animals of the British Isles: West Midlands by Dr Karl Shuker (2010)
The Mystery Animals of the British Isles: Dorset by Jonathan McGowan (2010)
The Mystery Animals of the British Isles: Sussex by Neil Arnold (2010)
The Mystery Animals of the British Isles: Devon by Jonathan Downes (2010)
The Mystery Animals of the British Isles: South Wales by Oll Lewis (2010)
The Mystery Animals of the British Isles: Mid Wales by Oll Lewis (2011)
The Mystery Animals of the British Isles: North Wales by Oll Lewis (TBA)
The Mystery Animals of the British Isles: Cornwall by Jon Downes (TBA)
The Mystery Animals of the British Isles: Yorkshire by Richard Freeman (TBA)

The Irish volumes will, at present, be divided up between Ronan Coghlan and Gary Cunningham, but - as yet - there are no dates or further information scheduled. If anyone reading this would like to volunteer to write any further volumes for this series, please get in touch with your suggestions.

OTHER BOOKS

Also during 2008 we published the following 'stand alone' titles:

- *In the Wake of Bernard Heuvelmans* by Karl Shuker and Michael A Woodley

Ever since humankind first ventured out onto the oceans, sailors came back with stories of sea monsters. For two hundred years, scientists have been attempting to classify these 'creatures' within an acceptable zoological frame of reference. The most important of these was produced by Professor Bernard Heuvelmans half a century ago. Michael Woodley, takes a look at Heuvelmans' classification model, re-examines it in the light of new discoveries in palaeontology and ichthyology over the past fifty years, and reaches some astounding conclusions.

- *The Island of Paradise: Chupacabra, UFO Crash Retrievals, and Accelerated Evolution on the Island of Puerto Rico* by Jonathan Downes

In his first book of original research for four years, Jon Downes visits the Antillean island of Puerto Rico, to which he has led two expeditions - in 1998 and 2004. Together with noted researcher Nick Redfern he goes in search of the grotesque vampiric chupacabra, believing that it can - finally - be categorised within a zoological frame of reference rather than a purely paranormal one. Along the way he uncovers mystery after mystery, has a run in with terrorists, art historians, and even has his garden buzzed by a UFO. By turns both terrifying and funny, this remarkable book is a real tour de force by one of the world's foremost cryptozoological researchers.

- *Dr Shuker's Casebook* by Dr Karl Shuker

Compiled here for the very first time, are some of the extraordinary cases that Karl has re-examined or personally explored down through the years - from sky beasts and reptoids, statues that weep, bleed, and even come to life, vanishing planets and invisible saints, frog rain and angel hair, and the world's weirdest ghosts and aliens, to a chiming tower of porcelain and a talking head of brass, spooklights and foo fighters, Herne the Hunter and photographed thought-forms, the chirping pyramid of Quetzalcoatl, magical mirrii dogs Down Under, and the most comprehensive study ever published of winged cats in which he successfully unveils their long-debated cryptic identity.

- *Dinosaurs and Other Prehistoric Animals on Stamps - A Worldwide Catalogue* by Karl P.N Shuker

This invaluable book will undoubtedly encourage everyone with a passion for dinosaurs and other prehistoric creatures to pursue it not only on screen, in books, or in museums but also via the ever-fascinating world of philately.

NEXT YEAR'S TITLES

Next year, as well as the 2010 Yearbook, and the volumes in the *Mystery Animals of the British Isles* series listed above, we plan to publish volumes four and five of our collected editions of *Animals & Men*, taking the series up to, and including #25 which was published in 2001. As a result of this, when current stocks of back issues of the magazine run out, issues 1-25 will not be reprinted in magazine form, although issues 26 onwards (we are currently at #45) will still remain available. This is purely because of lack of warehousing space.

We shall also be publishing a number of other volumes including:

Giant Snakes by Michael Newton
Tales from the CFZ by Nick Redfern
The Madness of Butterflies by Jonathan Downes
The Yellow Peril by Richard Muirhead
In Search of Prehistoric Survivors (updated) by Dr Karl Shuker
The Great Yokai Encyclopaedia - an A-Z of Japanese Monsters by Richard Freeman
Lytham and Booze by Tony `Doc` Shiels

Monstrum (new ed.) by Tony 'Doc' Shiels

We also intend to publish a series of books in conjunction with our magazine *Exotic Pets*, including books by Graham Smith and others.

RUSSIA EXPEDITION

In June, we sent a five-man expedition to the Russian republic of Kabardino-Balkaria in search of the fabled *almasty* or Russian Wildman. The expedition, which was partially funded by a generous donation from Professor Bryan Sykes, of Oxford Ancestors Ltd, and Wolfson College, Oxford, lasted for three weeks and was – to a certain extent at least – a success.

They returned with bone, hair and scat samples. Several sets of material have been sent off for analysis, and the first set of results have come back. Sadly, the hair samples, taken from a 'nest' high in the mountains, turned out to be of human origin.

We await the results of the rest of the sample analyses with interest. The book of the expedition was published in late November, and a feature length film is in preparation.

WEIRD WEEKEND 2008

In August we held our ninth annual convention, and the third to be held at Woolfardisworthy Community Centre. The following speakers appeared:

Matt Salusbury, Geoff Ward, Richard Freeman, Jonathan Downes, Ronan Coghlan, Dr Karl Shuker, Jon McGowan, Mike Hallowell, Dr Gail-Nina Anderson, Dr Mike Dash, Tim Matthews, Richard Ingram, Chris Moiser, Oll Lewis, Michael A Woodley. There were exhibitions from Rebecca McGowan-Griffin, Metamorphosis, and Ben Leese.

CFZ Award winners were: Mrs Marjorie Braund, Dr Karl Shuker, Adam Davies, Ronan Coghlan, Simon and Sharon Bennett

We would like to particularly thank David and Joanne Curtis, who not only came down from Co Durham at their own expense, but spent the whole weekend running a range of children's activities, which they financed themselves. Thank you, my dears, from the bottom of my heart.

WEIRD WEEKEND 2009

Next year's event will be held on 15-17 August 2009. Speakers confirmed so far include:

Nick Redfern, Max Blake, Andy Roberts, Richard Freeman, Jonathan Downes, Ronan Coghlan, Dr Karl Shuker, Jon McGowan, Tim Matthews, Oll Lewis, Michael A Woodley, Paul 'Mr Biffo' Rose, Neil Arnold, and more to be confirmed soon.

CFZ OUTREACH

Things are changing at the CFZ. We have to adapt to the times, but although our remit remains the same;

to study cryptozoology and allied disciplines, and to educate the public about these subjects, we are changing the way we do it. CFZ Press and CFZtv will remain unchanged. We are, however, launching two new branches, and making changes to a third.

- Museum Outreach

The CFZ Museum will remain in its current location, but we are rapidly becoming aware that we simply do not have enough space there to exhibit all that we want to. Whilst it would be a fantastic thing to have premises large enough for us to be able to site the whole museum under one roof, we accept that given the present state both of *our* finances, and of the global economy as a whole, this is unlikely to happen any time soon, if ever.

So we have decided to adopt a radical new approach. During 2009 we will start facilitating permanent and semi-permanent exhibitions in pubs, hotels, holiday centres, and tourist attractions across the region. This will allow us to invest in museum exhibits to a greater extent than we would otherwise have been able to, and will - we believe - act as a consciousness raising programme.

- Educational outreach

In the last six months we have had several encounters with young people, which have, frankly, horrified us. For example:

a. A 17 year old boy with eight GCSEs who was brought to me by his mother because "he wanted to be a zoologist". It turned out that he actually wanted to draw pokemon characters, but had seen an eagle at a falconry display and thought that it looked nice. He did not know the difference between reptiles, amphibians and fish.
b. A 13 year old who has won a scholarship to a public school who could not name any of the continents on a map, and did not know that most reptiles lay eggs, or that the thing that distinguishes mammals is that they produce milk.
c. A 20-something year old graduate who told me, pompously, that "It was a pity that I had not gone to university, or I would have known that these days China and Indo China are considered to be the same thing".
d. Another graduate who didn't know that St John's Gospel was in the New Testament, or that our present monarch was the daughter of King George VI. His excuse for this was that "he wasn't either a Christian or a Monarchist".
e. Two qualified (presumably) classroom assistants who thought that pumas and lynx were found in the UK.
f. A teenage girl studying sciences for GCSE who did not know what the binomial system of classification was, and had no idea what a chromosome was.
g. A teenage boy with 8 GCSEs who didn't know what I meant by 'New World' and thought that Columbus had gone on to discover Australia, and whose teacher had TOLD him that Cyrano de Bergerac was set in the Middle Ages.

We feel that the biggest cause of such lamentable levels of pure ignorance is an educational system that teaches kids to pass exams, but little else, and which certainly does nothing to foster a joy of knowledge for knowledge's sake. We intend to do a little bit to change that.

We are launching a community project with various community groups, which will feature kids, the long term unemployed *and* the disabled. We intend to visit schools and institutions with a multimedia roadshow, mixing theatre, film, science and art.

- Natural History Outreach

Natural History is no longer seen as a suitable hobby for young people, but in many cases is now an illegal one. A colleague of mine who works for the BBC Natural History Unit told me that children making a documentary on pond dipping were forced to wear safety helmets and rubber gloves before they were allowed near a garden pond. It is now illegal to take frogspawn from your own garden pond and put it in a fishtank. Changing social mores have meant that most of the children of people I know sit indoors all day playing computer games rather than exploring what little countryside is left. This may seem trivial to you, but Darwin, Linneaus, Mendel, and Gerald Durrell, amongst many others, were amateur naturalists first and foremost. Most professional zoologists started off as amateur naturalists. If kids are no longer able, or encouraged, to do this is it any wonder most of them seem to want to grow up to be image consultants or TV presenters?

We intend, again utilising kids, the long term unemployed *and* the disabled, to carry out a string of community projects over the next two years, aimed at raising children's interest in Natural History. This is something that we are already doing with *Exotic Pets* magazine, but we intend to expand our activities and target groups.

- Partnerships in Conservation

Over the last year of so, we have been travelling around the country looking at various zoos. And one thing that is becoming increasingly obvious is that most of them have exactly the same animals in them. The only real difference is that they are better exhibited in some zoos than others. This is - of course - a generalisation, but if you look hard enough you will see what we mean. Every zoo we have visited has at least three of the following species:

Bennett's wallaby
Parma wallaby
Coati
Meerkat
Ring-tailed lemur

This is largely because of the burgeoning amount of government (both UK, and European) legislation on the matter, as well as the code of conduct for members of BIAZA (The British and Irish Association of Zoos and Aquariums). Nobody would deny that the concept behind this legislation is sound enough. There is no justification, for example, for zoos selling their surplus stock for financial profit. But the legislation is - like so much Govermental action - completely ill thought out, and is going to lead to a complete stultification of the zoo establishment in the United Kingdom if something is not done about it.

Something that we find even more peculiar, is that although there are equally swingeing laws governing the private animal keepers,there are animals which are kept within the private sector which can never be seen in any zoo, despite the fact that they would make fine, and educational, exhibits.

There are animals in the CFZ museum collection, for example, which cannot be seen in any zoo, and really should be.
We have therefore come up with what we believe to be an innovative new idea, and are in the process of launching what we call 'The Partners in Conservation Initiative". We are actively seeking ongoing relationships with conservation and animal welfare organisations in the private sector.

In this way we can showcase the invaluable work done by these organisations, help to publicise their activities, and work together with them to educate the general public as to how they can get involved with research and conservation work around the world, before it is too late to save what is left of our planet We are at present in talks with several non-commercial organisations within the private sector, with the aim of launching the first couple of these partnership agreements before the beginning of the next season.

WHY THE OUTREACH PROGRAMME?

We believe that fortean zoology is a perfect catalyst to help the aims of both the Educational Outreach, and Natural History Outreach programmes. One of the reasons that we have spent so much of the last seventeen years trying to steer cryptozoology and the allied disciplines which make up the portmanteau discipline of 'Fortean Zoology' away from the cod-spirituality and New Age nonsense of the 'Mind, Body and Spirit' brigade, is that we believe that these subjects – especially cryptozoology in its purest form – deserve to be taken seriously as scientific disciplines.

To misquote the oft used line from *Hamlet* there IS more in heaven and earth than is dreamt of in the philosophy of much modern science, and certainly than is included in the cynical mishmash that is much of the National Curriculum. We believe that, although a great deal of what is written about cryptozoology, especially on the internet, is nonsense, a great deal isn't. We believe that through learning about myths and monsters, and how some of them may be true, and others most certainly are not, children can be taught not to blindly accept what they are told, but how to reason and use critical thinking.

We also believe that the search for mystery animals, even on a small scale, can kindle the fascination with the natural world, which is so sadly lacking in so much of modern youth. One hesitates to appear pretentious and quote Shakespeare twice in two paragraphs, but our ultimate aim is that the children with whom we work *"Find tongues in trees, books in running brooks, sermons in stones, and good in everything"*

THANK YOU

Thank you to everyone who has helped us through the last year. I would like to single out two in particular: Matthew Osborne and Dave Braund-Phillips, both of whom have worked tirelessly, and always with good humour, even when the hours they have been forced to work have been far beyond anything allowed under basic human rights legislation. Thank-you boys.

Finally, I would like to thank you all for your support during 2008, and look forward to your continued support during 2009. May you have a peaceful and happy holiday season, and the New Year that you would wish for yourselves.

God bless you all,

Jon Downes
(Director, CFZ)
Woolsery,
North Devon

December 5th 2008.

2008 in pictures

MAY: Finishing the museum

JUNE: Russia

MARCH: Big Cat Conference

AUGUST: Weird Weekend

THE CENTRE FOR FORTEAN ZOOLOGY

So, what is the Centre for Fortean Zoology?

We are a non profit-making organisation founded in 1992 with the aim of being a clearing house for information, and coordinating research into mystery animals around the world. We also study out of place animals, rare and aberrant animal behaviour, and Zooform Phenomena; little-understood "things" that appear to be animals, but which are in fact nothing of the sort, and not even alive (at least in the way we understand the term).

Why should I join the Centre for Fortean Zoology?

Not only are we the biggest organisation of our type in the world, but - or so we like to think - we are the best. We are certainly the only truly global Cryptozoological research organisation, and we carry out our investigations using a strictly scientific set of guidelines. We are expanding all the time and looking to recruit new members to help us in our research into mysterious animals and strange creatures across the globe. Why should you join us? Because, if you are genuinely interested in trying to solve the last great mysteries of Mother Nature, there is nobody better than us with whom to do it.

What do I get if I join the Centre for Fortean Zoology?

For £12 a year, you get a four-issue subscription to our journal *Animals & Men*. Each issue contains 60 pages packed with news, articles, letters, research papers, field reports, and even a gossip column! The magazine is A5 in format with a full colour cover. You also have access to one of the world's largest collections of resource material dealing with cryptozoology and allied disciplines, and people from the CFZ membership regularly take part in fieldwork and expeditions around the world.

How is the Centre for Fortean Zoology organized?

The CFZ is managed by a three-man board of trustees, with a non-profit making trust registered with HM Government Stamp Office. The board of trustees is supported by a Permanent Directorate of full and part-time staff, and advised by a Consultancy Board of specialists - many of whom who are world-renowned experts in their particular field. We have regional representatives across the UK, the USA, and many other parts of the world, and are affiliated with other organisations whose aims and protocols mirror our own.

I am new to the subject, and although I am interested I have little practical knowledge. I don't want to feel out of my depth. What should I do?

Don't worry. We were *all* beginners once. You'll find that the people at the CFZ are friendly and approachable. We have a thriving forum on the website which is the hub of an ever-growing electronic community. You will soon find your feet. Many members of the CFZ Permanent Directorate started off as ordinary members, and now work full-time chasing monsters around the world.

I have an idea for a project which isn't on your website. What do I do?

Write to us, e-mail us, or telephone us. The list of future projects on the website is not exhaustive. If you have a good idea for an investigation, please tell us. We may well be able to help.

How do I go on an expedition?

We are always looking for volunteers to join us. If you see a project that interests you, do not hesitate to get in touch with us. Under certain circumstances we can help provide funding for your trip. If you look on the future projects section of the website, you can see some of the projects that we have pencilled in for the next few years.

In 2003 and 2004 we sent three-man expeditions to Sumatra looking for Orang-Pendek - a semi-legendary bipedal ape. The same three went to Mongolia in 2005. All three members started off merely subscribers to the CFZ magazine.

Next time it could be you!

Project Kerinci, Sumatra - 2003
In search of the bipedal ape Orang Pendek

How is the Centre for Fortean Zoology funded?

We have no magic sources of income. All our funds come from donations, membership fees, works that we do for TV, radio or magazines, and sales of our publications and merchandise. We are always looking for corporate sponsorship, and other sources of revenue. If you have any ideas for fund-raising please let us know. However, unlike other cryptozoological organisations in the past, we do not live in an intellectual ivory tower. We are not afraid to get our hands dirty, and furthermore we are not one of those organisations where the membership have to raise money so that a privileged few can go on expensive foreign trips. Our research teams both in the UK and abroad, consist of a mixture of experienced and inexperienced personnel. We are truly a community, and work on the premise that the benefits of CFZ membership are open to all.

What do you do with the data you gather from your investigations and expeditions?

Reports of our investigations are published on our website as soon as they are available. Preliminary reports are posted within days of the project finishing.

Each year we publish a 200 page yearbook containing research papers and expedition reports too long to be printed in the journal. We freely circulate our information to anybody who asks for it.

Is the CFZ community purely an electronic one?

No. Each year since 2000 we have held our annual convention - the *Weird Weekend* - in Exeter. It is three days of lectures, workshops, and excursions. But most importantly it is a chance for members of the CFZ to meet each other, and to talk with the members of the permanent directorate in a relaxed and informal setting and preferably with a pint of beer in one hand. Since 2006 - the *Weird Weekend* has been bigger and better and held in the idyllic rural location of Woolsery in North Devon. The 2008 event will be held over the weekend 15-17 August.

Since relocating to North Devon in 2005 we have become ever more closely involved with other community organisations, and we hope that this trend will continue. We also work closely with Police Forces across the UK as consultants for animal mutilation cases, and we intend to forge closer links with the coastguard and other community services. We want to work closely with those who regularly travel into the Bristol Channel, so that if the recent trend of exotic animal visitors to our coastal waters continues, we can be out there as soon as possible.

We are building a Visitor's Centre in rural North Devon. This will not be open to the general public, but will provide a museum, a library and an educational resource for our members (currently over 400) across the globe. We are also planning a youth organisation which will involve children and young people in our activities. We work closely with *Tropiquaria* - a small zoo in north Somerset, and have several exciting conservation projects planned.

Apart from having been the only Fortean Zoological organisation in the world to have consistently published material on all aspects of the subject for over a decade, we have achieved the following concrete results:

- Disproved the myth relating to the headless so-called sea-serpent carcass of Durgan beach in Cornwall 1975
- Disproved the story of the 1988 puma skull of Lustleigh Cleave
- Carried out the only in-depth research ever into the mythos of the Cornish Owlman
- Made the first records of a tropical species of lamprey
- Made the first records of a luminous cave gnat larva in Thailand.
- Discovered a possible new species of British mammal - the beech marten.
- In 1994-6 carried out the first archival fortean zoological survey of Hong Kong.
- In the year 2000, CFZ theories where confirmed when an entirely new species of lizard was found resident in Britain.
- Identified the monster of Martin Mere in Lancashire as a giant wels catfish
- Expanded the known range of Armitage's skink in the Gambia by 80%
- Obtained photographic evidence of the remains of Europe's largest known pike
- Carried out the first ever in-depth study of the *ninki-nanka*
- Carried out the first attempt to breed Puerto Rican cave snails in captivity
- Were the first European explorers to visit the `lost valley` in Sumatra
- Published the first ever evidence for a new tribe of pygmies in Guyana
- Published the first evidence for a new species of caiman in Guyana

EXPEDITIONS & INVESTIGATIONS TO DATE INCLUDE:

- 1998 Puerto Rico, Florida, Mexico *(Chupacabras)*
- 1999 Nevada *(Bigfoot)*
- 2000 Thailand *(Giant snakes called nagas)*
- 2002 Martin Mere *(Giant catfish)*
- 2002 Cleveland *(Wallaby mutilation)*
- 2003 Bolam Lake *(BHM Reports)*
- 2003 Sumatra *(Orang Pendek)*
- 2003 Texas *(Bigfoot; giant snapping turtles)*
- 2004 Sumatra *(Orang Pendek; cigau, a sabre-toothed cat)*
- 2004 Illinois *(Black panthers; cicada swarm)*
- 2004 Texas *(Mystery blue dog)*
- 2004 Puerto Rico *(Chupacabras; carnivorous cave snails)*
- 2005 Belize *(Affiliate expedition for hairy dwarfs)*
- 2005 Mongolia *(Allghoi Khorkhoi aka Mongolian death worm)*
- 2006 Gambia *(Gambo - Gambian sea monster, Ninki Nanka and Armitage s skink*
- 2006 Llangorse Lake *(Giant pike, giant eels)*
- 2006 Windermere *(Giant eels)*
- 2007 Coniston Water *(Giant eels)*
- 2007 Guyana *(Giant anaconda, didi, water tiger)*
- 2008 Russia *(Almasty)*

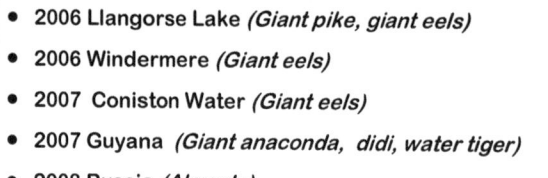

To apply for a <u>FREE</u> information pack about the organisation and details of how to join, plus information on current and future projects, expeditions and events.

Send a stamped and addressed envelope to:

**THE CENTRE FOR FORTEAN ZOOLOGY
MYRTLE COTTAGE, WOOLSERY,
BIDEFORD, NORTH DEVON
EX39 5QR.**

or alternatively visit our website at:
www.cfz.org.uk

Other books available from
CFZ PRESS

THE OWLMAN AND OTHERS - 30th Anniversary Edition
Jonathan Downes - ISBN 978-1-905723-02-7

£14.99

EASTER 1976 - Two young girls playing in the churchyard of Mawnan Old Church in southern Cornwall were frightened by what they described as a "nasty bird-man". A series of sightings that has continued to the present day. These grotesque and frightening episodes have fascinated researchers for three decades now, and one man has spent years collecting all the available evidence into a book. To mark the 30th anniversary of these sightings, Jonathan Downes has published a special edition of his book.

DRAGONS - More than a myth?
Richard Freeman - ISBN 0-9512872-9-X

£14.99

First scientific look at dragons since 1884. It looks at dragon legends worldwide, and examines modern sightings of dragon-like creatures, as well as some of the more esoteric theories surrounding dragonkind.

Dragons are discussed from a folkloric, historical and cryptozoological perspective, and Richard Freeman concludes that: "When your parents told you that dragons don't exist - they lied!"

MONSTER HUNTER
Jonathan Downes - ISBN 0-9512872-7-3

£14.99

Jonathan Downes' long-awaited autobiography, *Monster Hunter*...

Written with refreshing candour, it is the extraordinary story of an extraordinary life, in which the author crosses paths with wizards, rock stars, terrorists, and a bewildering array of mythical and not so mythical monsters, and still just about manages to emerge with his sanity intact.......

MONSTER OF THE MERE
Jonathan Downes - ISBN 0-9512872-2-2

£12.50

It all starts on Valentine's Day 2002 when a Lancashire newspaper announces that "Something" has been attacking swans at a nature reserve in Lancashire. Eyewitnesses have reported that a giant unknown creature has been dragging fully grown swans beneath the water at Martin Mere. An intrepid team from the Exeter based Centre for Fortean Zoology, led by the author, make two trips – each of a week – to the lake and its surrounding marshlands. During their investigations they uncover a thrilling and complex web of historical fact and fancy, quasi Fortean occurrences, strange animals and even human sacrifice.

**CFZ PRESS, MYRTLE COTTAGE,
WOOLFARDISWORTHY BIDEFORD,
NORTH DEVON, EX39 5QR
www.cfz.org.uk**

Other books available from
CFZ PRESS

ONLY FOOLS AND GOATSUCKERS
Jonathan Downes - ISBN 0-9512872-3-0

£12.50

In January and February 1998 Jonathan Downes and Graham Inglis of the Centre for Fortean Zoology spent three and a half weeks in Puerto Rico, Mexico and Florida, accompanied by a film crew from UK Channel 4 TV. Their aim was to make a documentary about the terrifying chupacabra - a vampiric creature that exists somewhere in the grey area between folklore and reality. This remarkable book tells the gripping, sometimes scary, and often hilariously funny story of how the boys from the CFZ did their best to subvert the medium of contemporary TV documentary making and actually do their job.

WHILE THE CAT'S AWAY
Chris Moiser - ISBN: 0-9512872-1-4

£7.99

Over the past thirty years or so there have been numerous sightings of large exotic cats, including black leopards, pumas and lynx, in the South West of England. Former Rhodesian soldier Sam McCall moved to North Devon and became a farmer and pub owner when Rhodesia became Zimbabwe in 1980. Over the years despite many of his pub regulars having seen the "Beast of Exmoor" Sam wasn't at all sure that it existed. Then a series of happenings made him change his mind. Chris Moiser—a zoologist—is well known for his research into the mystery cats of the westcountry. This is his first novel.

CFZ EXPEDITION REPORT 2006 - GAMBIA
ISBN 1905723032

£12.50

In July 2006, The J.T.Downes memorial Gambia Expedition - a six-person team - Chris Moiser, Richard Freeman, Chris Clarke, Oll Lewis, Lisa Dowley and Suzi Marsh went to the Gambia, West Africa. They went in search of a dragon-like creature, known to the natives as `Ninki Nanka`, which has terrorized the tiny African state for generations, and has reportedly killed people as recently as the 1990s. They also went to dig up part of a beach where an amateur naturalist claims to have buried the carcass of a mysterious fifteen foot sea monster named 'Gambo', and they sought to find the Armitage's Skink (*Chalcides armitagei*) - a tiny lizard first described in 1922 and only rediscovered in 1989. Here, for the first time, is their story.... With an forward by Dr. Karl Shuker and introduction by Jonathan Downes.

BIG CATS IN BRITAIN YEARBOOK 2006
Edited by Mark Fraser - ISBN 978-1905723-01-0

£10.00

Big cats are said to roam the British Isles and Ireland even now as you are sitting and reading this. People from all walks of life encounter these mysterious felines on a daily basis in every nook and cranny of these two countries. Most are jet-black, some are white, some are brown, in fact big cats of every description and colour are seen by some unsuspecting person while on his or her daily business. 'Big Cats in Britain' are the largest and most active group in the British Isles and Ireland This is their first book. It contains a run-down of every known big cat sighting in the UK during 2005, together with essays by various luminaries of the British big cat research community which place the phenomenon into scientific, cultural, and historical perspective.

CFZ PRESS, MYRTLE COTTAGE,
WOOLSERY, BIDEFORD,
NORTH DEVON, EX39 5QR
www.cfz.org.uk

Other books available from
CFZ PRESS

THE SMALLER MYSTERY CARNIVORES OF THE WESTCOUNTRY
Jonathan Downes - ISBN 978-1-905723-05-8

£7.99

Although much has been written in recent years about the mystery big cats which have been reported stalking Westcountry moorlands, little has been written on the subject of the smaller British mystery carnivores. This unique book redresses the balance and examines the current status in the Westcountry of three species thought to be extinct: the Wildcat, the Pine Marten and the Polecat, finding that the truth is far more exciting than the currently held scientific dogma. This book also uncovers evidence suggesting that even more exotic species of small mammal may lurk hitherto unsuspected in the countryside of Devon, Cornwall, Somerset and Dorset.

THE BLACKDOWN MYSTERY
Jonathan Downes - ISBN 978-1-905723-00-3

£7.99

Intrepid members of the CFZ are up to the challenge, and manage to entangle themselves thoroughly in the bizarre trappings of this case. This is the soft underbelly of ufology, rife with unsavoury characters, plenty of drugs and booze." That sums it up quite well, we think. A new edition of the classic 1999 book by legendary fortean author Jonathan Downes. In this remarkable book, Jon weaves a complex tale of conspiracy, anti-conspiracy, quasi-conspiracy and downright lies surrounding an air-crash and alleged UFO incident in Somerset during 1996. However the story is much stranger than that. This excellent and amusing book lifts the lid off much of contemporary forteana and explains far more than it initially promises.

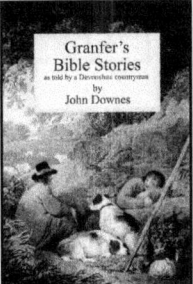

GRANFER'S BIBLE STORIES
John Downes - ISBN 0-9512872-8-1

£7.99

Bible stories in the Devonshire vernacular, each story being told by an old Devon Grandfather - 'Granfer'. These stories are now collected together in a remarkable book presenting selected parts of the Bible as one more-or-less continuous tale in short 'bite sized' stories intended for dipping into or even for bed-time reading. `Granfer` treats the biblical characters as if they were simple country folk living in the next village. Many of the stories are treated with a degree of bucolic humour and kindly irreverence, which not only gives the reader an opportunity to re-evaluate familiar tales in a new light, but do so in both an entertaining and a spiritually uplifting manner.

FRAGRANT HARBOURS DISTANT RIVERS
John Downes - ISBN 0-9512872-5-7

£12.50

Many excellent books have been written about Africa during the second half of the 19th Century, but this one is unique in that it presents the stories of a dozen different people, whose interlinked lives and achievements have as many nuances as any contemporary soap opera. It explains how the events in China and Hong Kong which surrounded the Opium Wars, intimately effected the events in Africa which take up the majority of this book. The author served in the Colonial Service in Nigeria and Hong Kong, during which he found himself following in the footsteps of one of the main characters in this book; Frederick Lugard – the architect of modern Nigeria.

**CFZ PRESS, MYRTLE COTTAGE,
WOOLFARDISWORTHY BIDEFORD,
NORTH DEVON, EX39 5QR
w w w . c f z . o r g . u k**

Other books available from
CFZ PRESS

ANIMALS & MEN - Issues 1 - 5 - In the Beginning
Edited by Jonathan Downes - ISBN 0-9512872-6-5

£12.50

At the beginning of the 21st Century monsters still roam the remote, and sometimes not so remote, corners of our planet. It is our job to search for them. The Centre for Fortean Zoology [CFZ] is the only professional, scientific and full-time organisation in the world dedicated to cryptozoology - the study of unknown animals. Since 1992 the CFZ has carried out an unparalleled programme of research and investigation all over the world. We have carried out expeditions to Sumatra (2003 and 2004), Mongolia (2005), Puerto Rico (1998 and 2004), Mexico (1998), Thailand (2000), Florida (1998), Nevada (1999 and 2003), Texas (2003 and 2004), and Illinois (2004). An introductory essay by Jonathan Downes, notes putting each issue into a historical perspective, and a history of the CFZ.

ANIMALS & MEN - Issues 6 - 10 - The Number of the Beast
Edited by Jonathan Downes - ISBN 978-1-905723-06-5

£12.50

At the beginning of the 21st Century monsters still roam the remote, and sometimes not so remote, corners of our planet. It is our job to search for them. The Centre for Fortean Zoology [CFZ] is the only professional, scientific and full-time organisation in the world dedicated to cryptozoology - the study of unknown animals. Since 1992 the CFZ has carried out an unparalleled programme of research and investigation all over the world. We have carried out expeditions to Sumatra (2003 and 2004), Mongolia (2005), Puerto Rico (1998 and 2004), Mexico (1998), Thailand (2000), Florida (1998), Nevada (1999 and 2003), Texas (2003 and 2004), and Illinois (2004). Preface by Mark North and an introductory essay by Jonathan Downes, notes putting each issue into a historical perspective, and a history of the CFZ.

BIG BIRD! Modern Sightings of Flying Monsters

Ken Gerhard - ISBN 978-1-905723-08-9

£7.99

From all over the dusty U.S./Mexican border come hair-raising stories of modern day encounters with winged monsters of immense size and terrifying appearance. Further field sightings of similar creatures are recorded from all around the globe. What lies behind these weird tales? Ken Gerhard is a native Texan, he lives in the homeland of the monster some call 'Big Bird'. Ken's scholarly work is the first of its kind. On the track of the monster, Ken uncovers cases of animal mutilations, attacks on humans and mounting evidence of a stunning zoological discovery ignored by mainstream science. Keep watching the skies!

STRENGTH THROUGH KOI
They saved Hitler's Koi and other stories

£7.99

Jonathan Downes - ISBN 978-1-905723-04-1

Strength through Koi is a book of short stories - some of them true, some of them less so - by noted cryptozoologist and raconteur Jonathan Downes. The stories are all about koi carp, and their interaction with bigfoot, UFOs, and Nazis. Even the late George Harrison makes an appearance. Very funny in parts, this book is highly recommended for anyone with even a passing interest in aquaculture, but should be taken definitely *cum grano salis*.

CFZ PRESS, MYRTLE COTTAGE, WOOLSERY, BIDEFORD, NORTH DEVON, EX39 5QR

Other books available from
CFZ PRESS

BIG CATS IN BRITAIN YEARBOOK 2007
Edited by Mark Fraser - ISBN 978-1-905723-09-6

£12.50

People from all walks of life encounter mysterious felids on a daily basis, in every nook and cranny of the UK. Most are jet-black, some are white, some are brown; big cats of every description and colour are seen by some unsuspecting person while on his or her daily business. 'Big Cats in Britain' are the largest and most active research group in the British Isles and Ireland. This book contains a run-down of every known big cat sighting in the UK during 2006, together with essays by various luminaries of the British big cat research community.

CAT FLAPS! Northern Mystery Cats
Andy Roberts - ISBN 978-1-905723-11-9

£6.99

Of all Britain's mystery beasts, the alien big cats are the most renowned. In recent years the notoriety of these uncatchable, out-of-place predators have eclipsed even the Loch Ness Monster. They slink from the shadows to terrorise a community, and then, as often as not, vanish like ghosts. But now film, photographs, livestock kills, and paw prints show that we can no longer deny the existence of these once-legendary beasts. Here then is a case-study, a true lost classic of Fortean research by one of the country's most respected researchers.

CENTRE FOR FORTEAN ZOOLOGY 2007 YEARBOOK
Edited by Jonathan Downes and Richard Freeman
ISBN 978-1-905723-14-0

£12.50

The Centre For Fortean Zoology Yearbook is a collection of papers and essays too long and detailed for publication in the CFZ Journal *Animals & Men*. With contributions from both well-known researchers, and relative newcomers to the field, the Yearbook provides a forum where new theories can be expounded, and work on little-known cryptids discussed.

MONSTER! THE A-Z OF ZOOFORM PHENOMENA
Neil Arnold - ISBN 978-1-905723-10-2

£14.99

Zooform Phenomena are the most elusive, and least understood, mystery `animals`. Indeed, they are not animals at all, and are not even animate in the accepted terms of the word. Author and researcher Neil Arnold is to be commended for a groundbreaking piece of work, and has provided the world's first alphabetical listing of zooforms from around the world.

**CFZ PRESS, MYRTLE COTTAGE,
WOOLFARDISWORTHY BIDEFORD,
NORTH DEVON, EX39 5QR
w w w . c f z . o r g . u k**

Other books available from
CFZ PRESS

BIG CATS LOOSE IN BRITAIN
Marcus Matthews - ISBN 978-1-905723-12-6

£14.99

Big Cats: Loose in Britain, looks at the body of anecdotal evidence for such creatures: sightings, livestock kills, paw-prints and photographs, and seeks to determine underlying commonalities and threads of evidence. These two strands are repeatedly woven together into a highly readable, yet scientifically compelling, overview of the big cat phenomenon in Britain.

DARK DORSET
TALES OF MYSTERY, WONDER AND TERROR
Robert. J. Newland and Mark. J. North
ISBN 978-1-905723-15-6

£12.50

This extensively illustrated compendium has over 400 tales and references, making this book by far one of the best in its field. Dark Dorset has been thoroughly researched, and includes many new entries and up to date information never before published. The title of the book speaks for itself, and is indeed not for the faint hearted or those easily shocked.

MAN-MONKEY - IN SEARCH OF THE BRITISH BIGFOOT
Nick Redfern - ISBN 978-1-905723-16-4

£9.99

In her 1883 book, *Shropshire Folklore*, Charlotte S. Burne wrote: *'Just before he reached the canal bridge, a strange black creature with great white eyes sprang out of the plantation by the roadside and alighted on his horse's back'*. The creature duly became known as the `Man-Monkey`.

Between 1986 and early 2001, Nick Redfern delved deeply into the mystery of the strange creature of that dark stretch of canal. Now, published for the very first time, are Nick's original interview notes, his files and discoveries; as well as his theories pertaining to what lies at the heart of this diabolical legend.

EXTRAORDINARY ANIMALS REVISITED
Dr Karl Shuker - ISBN 978-1905723171

£14.99

This delightful book is the long-awaited, greatly-expanded new edition of one of Dr Karl Shuker's much-loved early volumes, *Extraordinary Animals Worldwide*. It is a fascinating celebration of what used to be called romantic natural history, examining a dazzling diversity of animal anomalies, creatures of cryptozoology, and all manner of other thought-provoking zoological revelations and continuing controversies down through the ages of wildlife discovery.

CFZ PRESS, MYRTLE COTTAGE, WOOLFARDISWORTHY BIDEFORD, NORTH DEVON, EX39 5QR
w w w . c f z . o r g . u k

Other books available from
CFZ PRESS

DARK DORSET CALENDAR CUSTOMS
Robert J Newland - ISBN 978-1-905723-18-8

£12.50

Much of the intrinsic charm of Dorset folklore is owed to the importance of folk customs. Today only a small amount of these curious and occasionally eccentric customs have survived, while those that still continue have, for many of us, lost their original significance. Why do we eat pancakes on Shrove Tuesday? Why do children dance around the maypole on May Day? Why do we carve pumpkin lanterns at Hallowe'en? All the answers are here! Robert has made an in-depth study of the Dorset country calendar identifying the major feast-days, holidays and celebrations when traditionally such folk customs are practiced.

CENTRE FOR FORTEAN ZOOLOGY 2004 YEARBOOK
Edited by Jonathan Downes and Richard Freeman
ISBN 978-1-905723-14-0

£12.50

The Centre For Fortean Zoology Yearbook is a collection of papers and essays too long and detailed for publication in the CFZ Journal *Animals & Men*. With contributions from both well-known researchers, and relative newcomers to the field, the Yearbook provides a forum where new theories can be expounded, and work on little-known cryptids discussed.

CENTRE FOR FORTEAN ZOOLOGY 2008 YEARBOOK
Edited by Jonathan Downes and Corinna Downes
ISBN 978 -1-905723-19-5

£12.50

The Centre For Fortean Zoology Yearbook is a collection of papers and essays too long and detailed for publication in the CFZ Journal *Animals & Men*. With contributions from both well-known researchers, and relative newcomers to the field, the Yearbook provides a forum where new theories can be expounded, and work on little-known cryptids discussed.

ETHNA'S JOURNAL
Corinna Newton Downes
ISBN 978 -1-905723-21-8

£9.99

Ethna's Journal tells the story of a few months in an alternate Dark Ages, seen through the eyes of Ethna, daughter of Lord Edric. She is an unsophisticated girl from the fortress town of Cragnuth, somewhere in the north of England, who reluctantly gets embroiled in a web of treachery, sorcery and bloody war...

**CFZ PRESS, MYRTLE COTTAGE,
WOOLFARDISWORTHY BIDEFORD,
NORTH DEVON, EX39 5QR
www.cfz.org.uk**

Other books available from
CFZ PRESS

ANIMALS & MEN - Issues 11 - 15 - The Call of the Wild
Jonathan Downes (Ed) - ISBN 978-1-905723-07-2

£12.50

Since 1994 we have been publishing the world's only dedicated cryptozoology magazine, *Animals & Men*. This volume contains fascimile reprints of issues 11 to 15 and includes articles covering out of place walruses, feathered dinosaurs, possible North American ground sloth survival, the theory of initial bipedalism, mystery whales, mitten crabs in Britain, Barbary lions, out of place animals in Germany, mystery pangolins, the barking beast of Bath, Yorkshire ABCs, Molly the singing oyster, singing mice, the dragons of Yorkshire, singing mice, the bigfoot murders, waspman, British beavers, the migo, Nessie, the weird warbling whatsit of the westcountry, the quagga project and much more...

IN THE WAKE OF BERNARD HEUVELMANS
Michael A Woodley - ISBN 978-1-905723-20-1

£9.99

Everyone is familiar with the nautical maps from the middle ages that were liberally festooned with images of exotic and monstrous animals, but the truth of the matter is that the *idea* of the sea monster is probably as old as humankind itself.

For two hundred years, scientists have been producing speculative classifications of sea serpents, attempting to place them within a zoological framework. This book looks at these successive classification models, and using a new formula produces a sea serpent classification for the 21st Century.

CENTRE FOR FORTEAN ZOOLOGY 1999 YEARBOOK
Edited by Jonathan Downes
ISBN 978-1-905723-24-9

£12.50

The Centre For Fortean Zoology Yearbook is a collection of papers and essays too long and detailed for publication in the CFZ Journal *Animals & Men*. With contributions from both well-known researchers, and relative newcomers to the field, the Yearbook provides a forum where new theories can be expounded, and work on little-known cryptids discussed.

CENTRE FOR FORTEAN ZOOLOGY 1996 YEARBOOK
Edited by Jonathan Downes
ISBN 978-1-905723-22-5

£12.50

The Centre For Fortean Zoology Yearbook is a collection of papers and essays too long and detailed for publication in the CFZ Journal *Animals & Men*. With contributions from both well-known researchers, and relative newcomers to the field, the Yearbook provides a forum where new theories can be expounded, and work on little-known cryptids discussed.

**CFZ PRESS, MYRTLE COTTAGE,
WOOLFARDISWORTHY BIDEFORD,
NORTH DEVON, EX39 5QR
w w w . c f z . o r g . u k**

Other books available from
CFZ PRESS

BIG CATS IN BRITAIN YEARBOOK 2008
Edited by Mark Fraser - ISBN 978-1-905723-23-2

£12.50

People from all walks of life encounter mysterious felids on a daily basis, in every nook and cranny of the UK. Most are jet-black, some are white, some are brown; big cats of every description and colour are seen by some unsuspecting person while on his or her daily business. 'Big Cats in Britain' are the largest and most active research group in the British Isles and Ireland. This book contains a run-down of every known big cat sighting in the UK during 2007, together with essays by various luminaries of the British big cat research community.

CFZ EXPEDITION REPORT 2007 - GUYANA
ISBN 978-1-905723-25-6

£12.50

Since 1992, the CFZ has carried out an unparalleled programme of research and investigation all over the world. In November 2007, a five-person team - Richard Freeman, Chris Clarke, Paul Rose, Lisa Dowley and Jon Hare went to Guyana, South America. They went in search of giant anacondas, the bigfoot-like didi, and the terrifying water tiger.

Here, for the first time, is their story...With an introduction by Jonathan Downes and forward by Dr. Karl Shuker.

CENTRE FOR FORTEAN ZOOLOGY 2003 YEARBOOK
Edited by Jonathan Downes and Richard Freeman
ISBN 978 -1-905723-19-5

£12.50

The Centre For Fortean Zoology Yearbook is a collection of papers and essays too long and detailed for publication in the CFZ Journal *Animals & Men*. With contributions from both well-known researchers, and relative newcomers to the field, the Yearbook provides a forum where new theories can be expounded, and work on little-known cryptids discussed.

CENTRE FOR FORTEAN ZOOLOGY 1997 YEARBOOK
Edited by Jonathan Downes and Graham Inglis
ISBN 978 -1-905723-27-0

£12.50

The Centre For Fortean Zoology Yearbook is a collection of papers and essays too long and detailed for publication in the CFZ Journal *Animals & Men*. With contributions from both well-known researchers, and relative newcomers to the field, the Yearbook provides a forum where new theories can be expounded, and work on little-known cryptids discussed.

**CFZ PRESS, MYRTLE COTTAGE,
WOOLFARDISWORTHY BIDEFORD,
NORTH DEVON, EX39 5QR
w w w . c f z . o r g . u k**

Other books available from
CFZ PRESS

CENTRE FOR FORTEAN ZOOLOGY 2000-1 YEARBOOK
Edited by Jonathan Downes and Richard Freeman
ISBN 978-1-905723-19-5

£12.50

The Centre For Fortean Zoology Yearbook is a collection of papers and essays too long and detailed for publication in the CFZ Journal *Animals & Men*. With contributions from both well-known researchers, and relative newcomers to the field, the Yearbook provides a forum where new theories can be expounded, and work on little-known cryptids discussed.

THE MYSTERY ANIMALS OF THE BRITISH ISLES: NORTHUMBERLAND AND TYNESIDE
Michael J Hallowell
ISBN 978-1-905723-29-4

£12.50

Mystery animals? Great Britain? Surely not. But is is true.

This is a major new series from CFZ Press. It will cover Great Britain and the Republic of Ireland, on a county by county basis, describing the mystery animals of the entire island group.

The Island of Paradise: Chupacabra, UFO Crash Retrievals, and Accelerated Evolution on the Island of Puerto Rico
Jonathan Downes - ISBN 978-1-905723-32-4

£14.99

In his first book of original research for four years, Jon Downes visits the Antillean island of Puerto Rico, to which he has led two expeditions - in 1998 and 2004. Together with noted researcher Nick Redfern he goes in search of the grotesque vampiric chupacabra, believing that it can - finally - be categorised within a zoological frame of reference rather than a purely paranormal one. Along the way he uncovers mystery after mystery, has a run in with terrorists, art historians, and even has his garden buzzed by a UFO. By turns both terrifying and funny, this remarkable book is a real tour de force by one of the world's foremost cryptozoological researchers.

DR SHUKER'S CASEBOOK
Dr Karl Shuker - ISBN 978-1905723-33-1

£14.99

Although he is best-known for his extensive cryptozoological researches and publications, Dr Karl Shuker has also investigated a very diverse range of other anomalies and unexplained phenomena, both in the literature and in the field. Now, compiled here for the very first time, are some of the extraordinary cases that he has re-examined or personally explored down through the years.

**CFZ PRESS, MYRTLE COTTAGE,
WOOLFARDISWORTHY BIDEFORD,
NORTH DEVON, EX39 5QR
w w w . c f z . o r g . u k**

Other books available from
CFZ PRESS

Dinosaurs and Other Prehistoric Animals on Stamps: A Worldwide Catalogue
Dr Karl P.N.Shuker - ISBN 978-1-905723-34-8

£9.99

Compiled by zoologist Dr Karl P.N. Shuker, a lifelong, enthusiastic collector of wildlife stamps and with an especial interest in those that portray fossil species, it provides an exhaustive, definitive listing of stamps and miniature sheets depicting dinosaurs and other prehistoric animals issued by countries throughout the world. It also includes sections dealing with cryptozoological stamps, dinosaur stamp superlatives, and unofficial prehistoric animal stamps.

CENTRE FOR FORTEAN ZOOLOGY 2009 YEARBOOK
Edited by Jonathan Downes and Richard Freeman
ISBN 978 -1-905723-37-9

£12.50

The Centre For Fortean Zoology Yearbook is a collection of papers and essays too long and detailed for publication in the CFZ Journal *Animals & Men*. With contributions from both well-known researchers, and relative newcomers to the field, the Yearbook provides a forum where new theories can be expounded, and work on little-known cryptids discussed.

**CFZ PRESS, MYRTLE COTTAGE,
WOOLFARDISWORTHY BIDEFORD,
NORTH DEVON, EX39 5QR**
www.cfz.org.uk

www.ingramcontent.com/pod-product-compliance
Ingram Content Group UK Ltd.
Pitfield, Milton Keynes, MK11 3LW, UK
UKHW021319180426
11947UKWH00015B/1326